Advanced Organic Synthesis

A Laboratory Manual

Advanced Organic Synthesis

A Laboratory Manual

Dmitry V. Liskin

Christopher Newport University
Newport News, Virginia, USA

Penny Chaloner

Imperial College London
London, England

CRC Press
Taylor & Francis Group
Boca Raton London New York

CRC Press is an imprint of the
Taylor & Francis Group, an **informa** business

CRC Press
Taylor & Francis Group
6000 Broken Sound Parkway NW, Suite 300
Boca Raton, FL 33487-2742

Printed on acid-free paper
Version Date: 20151008

International Standard Book Number-13: 978-1-4822-4496-0 (Paperback)

Library of Congress Cataloging-in-Publication Data

Liskin, Dmitry V.
    Advanced organic synthesis : a laboratory manual / Dmitry V. Liskin and Penny Chaloner.
        pages cm
    "A CRC title."
    Includes bibliographical references and index.
    ISBN 978-1-4822-4496-0
    1. Organic compounds--Synthesis--Laboratory manuals. 2. Chemistry, Organic--Laboratory manuals. I. Chaloner, Penny A. II. Title.

QD262.L57 2016
547'.2078--dc23   2015027184

Visit the Taylor & Francis Web site at
http://www.taylorandfrancis.com

and the CRC Press Web site at
http://www.crcpress.com

To my parents Vladimir and Lyudmila and
my wife Kellie and daughter Mathilde, for their
continuing encouragement and support.

Contents

Preface

A few years ago, I was looking to develop an "Advanced Organic Lab" course in order to provide students with insights into graduate-level laboratory, equipment, and techniques. I wanted to teach what I wished I knew before starting graduate school. The purpose of this manual is just to provide such information and experience, focusing on *techniques* and *approaches* used in more advanced labs. This manual features some experiments that I performed while being a graduate student and recent experiments published in peer-reviewed journals and provides experience that will help you thrive in a lab.

This manual would have not been possible without help from the Taylor & Francis Group, especially Hilary LaFoe and Ashley Weinstein. I would also like to thank Dr. Penny Chaloner for her immense help, patience, and insight. Last, but not least, I would like to thank my dear friend Alex Novak for his encouragement.

Dmitry V. Liskin
Christopher Newport University
Newport News, Virginia

Authors

Dmitry V. Liskin, PhD, is a lecturer at Christopher Newport University, Newport News, Virginia. He graduated with a degree in ACS biochemistry from Mississippi College, Clinton, Mississippi. As an undergraduate, he worked on the synthesis and characterization of pseudoacids under the mentorship of Dr. Edward Valente and completed his doctorate studies in organic chemistry at the University of Washington in Seattle. Dr. Liskin joined a research group of Dr. Forrest Michael and developed alkene difunctionalizations as novel routes to substituted nitrogen heterocycles, mainly focusing on the use of hypervalent iodine oxidants.

Penny Chaloner, PhD, completed her undergraduate and graduate studies in Cambridge and had postdoctoral fellowships in Oxford, followed by a period in the United States as an assistant professor at Rutgers University and teaching at Harvard Summer School. She returned from the United States to a permanent position in Sussex, from which she recently retired. She first taught organic sophomore chemistry at Harvard in 1981 and has taught some version of this course once or twice a year ever since. Some of that was in North America, much in the United Kingdom, at Sussex, where for many years she taught the American organic chemistry sequence to visiting students and to a large group of international premeds. Penny Chaloner has experience teaching this sequence in groups sized from two to almost a thousand.

Introduction: Lab Setup, Safety, and Strategies for Efficient Working

I.1 Lab Setup

The laboratory of a synthetic chemist may look overwhelming and even frightening to a new student. There are definitely many more lines, clamps, tubes, complicated glassware, and, most importantly, scribbles on the hood sash, something not commonly encountered in a sophomore organic chemistry lab (Figure I.1).

Every piece of equipment has its own purpose. Manifolds, vacuum lines, distillation apparatuses, etc., may look slightly different from lab to lab, but they all serve the same purpose. The objective of this manual is to illustrate techniques and demonstrate safe and efficient ways of performing organic synthesis.

The average hood of a synthetic chemist is equipped with a glass manifold (Figure I.2) able to pull vacuum or deliver inert gas when the relevant tap is turned. This is probably the most prized possession as it is both costly and useful. The manifold must be kept clean and well greased in order to fulfill its functions as designed. It is always a good idea to calibrate your

Figure I.1 Typical research lab hood.

Figure I.2 Glass manifold with a cold trap.

manifold in order to know how much vacuum is being supplied. When removing volatile solvents under vacuum, ensure that the trap is filled with liquid nitrogen to guarantee a complete condensation of the volatile material—this prolongs the vacuum pump's life and makes servicing less frequent.

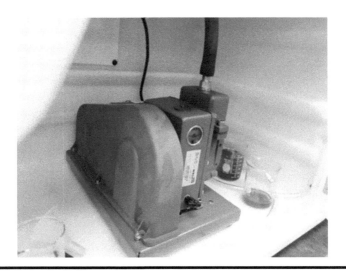

Figure I.3   Vacuum pump.

The vacuum is supplied by a vacuum pump, which usually lives under the hood (Figure I.3). Make sure that the oil is changed at prescribed intervals and that its level is optimal. Also, listen carefully to the sound it makes: A happy pump is a quiet one. It is normal to hear gurgling sounds for a few seconds when the tap is opened; however, prolonged gurgling is indicative of a vacuum leak.

The next important piece of equipment is the heating plate, usually combined with a stirring option (Figure I.4). This is a must in any lab hood. Its purpose is quite straightforward—to supply heat and stirring. It will take a bit of practice to learn which setting will provide a particular temperature. Sometimes these plates are coupled with a digital thermometer and can be set to sustain a desired temperature automatically. While extremely convenient, this option is rarely seen in undergraduate labs. When using the stirring option, ensure that the flask is as close as possible to the plate (depending on the size of the stirrer bar, stirring is possible with the reaction flask being a few centimeters off the plate) and, most importantly, centered. Following these guidelines will result in smooth and uninterrupted stirring.

Figure I.4   Stirrer hotplate on a jack stand.

Finally, there is a (usually metal) scaffold located toward the back wall of the hood. Its purpose is to provide support to numerous clamps. Every student will organize it according to his or her needs. In the absence of a scaffold, several ring stands will perform the same function.

The rest of the hood will be populated by stock solutions, various beakers and flasks. When organizing your glassware, ensure that all volatile and flammable reagents are kept away from sources of heat and/or open flames. Also, make sure that the electrical cords of the stirrer hotplate or any other piece of electrical equipment are placed behind the scaffold and along the back of the hood. This way there is less chance of knocking anything over with the cord. Follow same logic when running water/gas lines.

I.2 Safety

Working with chemicals, even in a well-equipped lab, poses potential health risks to scientists. Certainly, spills and other accidents are unavoidable, but a good student needs to know how to respond quickly to these emergencies and should practice techniques that will minimize chances of such accidents.

First off, always wear goggles! Your eyes may be permanently damaged by common chemicals. Wearing a lab coat is also necessary. Make sure that the one you choose complies with your university's code. Some experiments will *require* the use of gloves—there are various types available, and this is something you should take advice on if you are unsure.

This brings us to the next point: Always know the chemicals that you will be working with. Carefully check the material safety data sheet (MSDS) beforehand; know their reactivity, toxicity, and first aid in case of exposure. Follow commonsense practices. Keep flammables away from potential sources of ignition. Water-sensitive reagents must not only be kept away from bulk water, but their exposure to the atmosphere should also be minimized. If you have stock solutions of acids and bases, keep them separate.

Finally, know where the closest fire extinguisher, eye wash station, safety shower, and telephone are located. In case of any accident, however minor, immediately inform your instructor!

I.3 Efficiency Strategies

I.3.1 Have a Game Plan

Each chapter in this book contains a brief background of each experiment, step-by-step instructions, and a list of reagents and materials. Always read and understand the procedure before coming to the lab. Many times I have witnessed students finishing one or two hours earlier than their peers while performing the same experiment. Why does this happen? The answer is preparation and game plan. Students who come to lab prepared are able to complete their assignments early. Sure, that requires more time beforehand. However, the amount of time invested is well justified by the time gained by finishing early. Certainly, there is a learning curve: It is best to invest a bit more time at the beginning and to learn to do things right the first time, rather than trying to correct errors later.

I.3.2 Understand the Steps and Be Organized

Simply reading and outlining the reaction procedure is not enough to be really efficient. Each student must think critically and understand the purpose behind each step and technique. Know beforehand what reagents and glassware are needed and get them ready before starting the reaction. Understanding the difference between qualitative and quantitative steps is a major time saver. For example, when directions call to acidify the reaction mixture with 10 mL acid, measuring out 10.00 mL is absolutely unnecessary; 9.5–10.5 mL will do just fine. Chances are that more acid will be needed to be added based on a litmus test anyway. There are times when you need exactly 10.000 g of a reagent, and it is important to identify those steps. There is a big difference between cutting corners and being efficient. The first one never fails to produce sloppy results, while the second yields quality work in a fraction of time.

I.3.3 Stagger Your Tasks and Think Ahead

These are perhaps among the hardest things to learn, and managing your time efficiently comes with experience. The rewards, however, are immense. Many reactions will need some time to reflux, stir, cool, etc., in order for that chemistry magic to happen. In a research lab, chemists will never be found *sitting and waiting* for a one-hour reaction to complete—they will find other ways to be busy. Just as in a good game of chess, think three to four steps ahead, identify potential wait periods and think how you can be meaningfully occupied. Whether it is setting up another reaction or prepping for the next step, it is your time, so don't waste it. In reality, you will probably not need to set up a new reaction in an undergrad lab, but getting a water bath ready for the next step or spotting a TLC plate, cleaning dishes, or even doing your homework for another class is a better use of time than playing Tetris on your phone. If you put conscious effort into thinking about staggering your tasks and thinking ahead, you will very soon gain enough experience and intuition that you do not need to think about it at all.

Chapter 1

Diastereoselective Reduction of Estrone

Keywords: Reduction, diastereoselectivity, TLC

Safety: Sulfuric acid is very corrosive, reacts violently with water, and causes severe burns. Acetone, hexanes, and ethanol are volatile and flammable and should be handled in the hood. Dichloromethane is volatile and a known carcinogen and irritant. Sodium borohydride is toxic on contact and will emit flammable gases when reacted with water. Estrone and estradiol are suspected carcinogens. For full MSDS, visit www.sigmaaldrich.com.

Equipment Required

- Balance
- Oven
- Stirrer hotplate
- Thermometer
- Hirsch funnel and filter flask
- Sand bath or oil bath
- 25 mL round bottomed flask with a septum and stirrer bar

- Small (1–2 mL) syringe and needle
- Reflux condenser
- Heat gun
- Melting point apparatus and capillaries
- Dessicator
- TLC plates and developing chamber (200 mL beaker and tin foil will suffice)
- Forceps
- Pasteur pipettes
- Small (1 and 5 mL) vials with caps
- TLC elution mixture (prepare beforehand; 100 mL should be enough for the whole lab): acetone:dichloromethane:hexanes (22:50:28)
- Vanillin TLC developing solution (prepared by instructor): dissolve 15 g of vanillin in 250 mL of 95% ethanol and 2.5 mL of concentrated sulfuric acid. The solution should be stored in a capped container in a cool and dark place and discarded when the solution turns blue.

Materials Required

- Estrone
- 17-β-estradiol
- 17-α-estradiol
- $NaBH_4$
- Vanillin
- Acetone
- Dichloromethane
- Hexanes
- H_2SO_4, concentrated
- Ethanol

Purpose: To diastereoselectively reduce estrone and monitor the progress of the reaction by TLC.

Overall Reaction

Stereoselective reductions are important reactions that provide access to stereo-enriched materials (Scheme 1.1). In order to achieve stereoselectivity, at least one of the reagents has to be stereopure. In this lab, estrone, a steroid, will be reduced with sodium borohydride. Ketones are readily reduced by mild hydride donors, such as sodium borohydride, to yield alcohols. As Scheme 1.2 shows, there are two ways a hydride can attack this ketone.

Hydride attack from the *beta* (top) face produces the *alpha* alcohol and attack from the *alpha* (bottom) face gives the *beta* alcohol. The methyl group on estrone sterically hinders the *beta* face, rendering hydride attack from the top challenging. Hence, the major product is a more sterically hindered

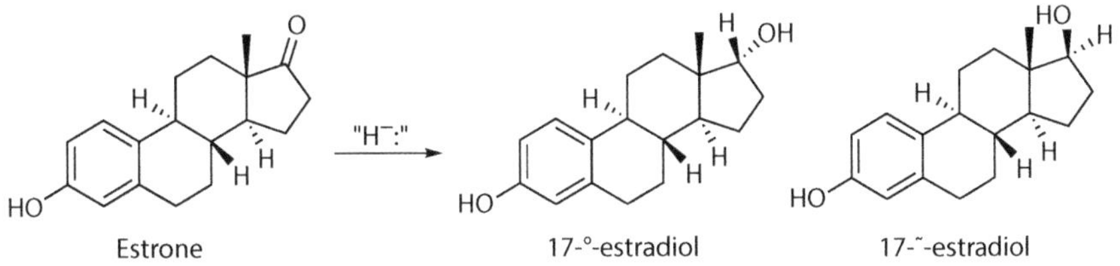

"H⁻:"
Estrone
17-°-estradiol
17-˝-estradiol

Scheme 1.1 Reduction of estrone.

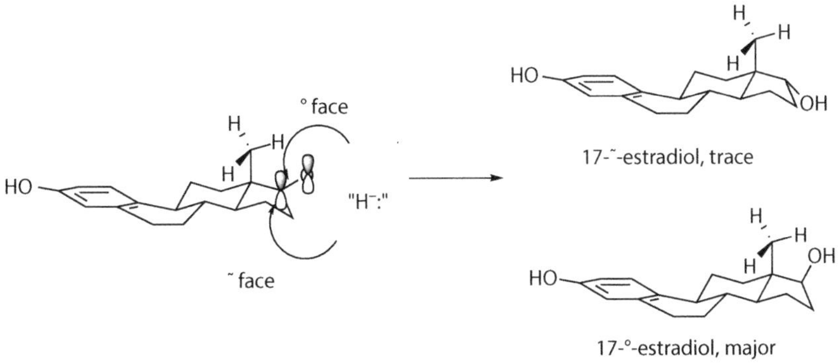

° face
˝ face
"H⁻:"
17-˝-estradiol, trace
17-°-estradiol, major

Scheme 1.2 Diastereoselective reduction of estrone.

product, resulting from attack on the more accessible side. This can be referred to as *steric approach control.*

Besides leading to an interesting stereochemical outcome, the main focus of this experiment is analyzing the progression of the reaction and its outcome. Using thin layer chromatography (TLC) is very common in organic synthesis as this method does not interrupt the reaction, and loss of product is negligible. It is also very quick and efficient and does not require sophisticated instruments. Many experiments in this manual will use TLC to monitor reaction progress.

A sample of the reaction mixture will be drawn by a thin spotter and placed on a TLC plate. Known solutions of starting material and both product stereoisomers will be added to the plate. Once the TLC is run and developed, observations regarding reaction progress and outcome will be assessed. You will be looking for disappearance of the estrone starting material and the appearance of the product(s). The distance traveled by the spot from the reaction should match one of the standards.

Once the product has been identified, it will be important to verify its identity by co-spotting. In this technique, TLC plate will be spotted with of crude reaction, suspected product, and a combination of both spots—three spots in total. After running and developing the plate, all three spots should have traveled the same distance.

Procedure

Lab Period 1

1. Prepare a fresh 0.5 M solution of $NaBH_4$ in absolute ethanol. In a pre-dried 5 mL vial, dissolve 56.7 mg of sodium borohydride (1.5 mmol) in 3 mL of absolute ethanol and cap immediately to avoid introduction of atmospheric water.

2. Equip a dry 25 mL round-bottomed flask with a stirrer bar and add 40 mg (0.14 mmol) of estrone and 5 mL of ethanol; cap the vial with a septum and place a syringe needle through the septum. As the reaction will be heated, the needle will act as a pressure release outlet.

3. Heat the reaction mixture to 70°C and stir until the estrone dissolves completely.*

4. Once the estrone is dissolved, draw 1 mL of borohydride solution into syringe and add to the reaction mixture.

5. Replace the septum with a condenser and continue heating and stirring for 30 min. Make sure that the temperature does not exceed 80°C.

6. Meanwhile, pull a few TLC spotters from Pasteur pipettes (see "Essential Laboratory Techniques") and prepare standard solutions of estrone and both estradiols. To prepare those solutions, place 5–7 mg of estrone, 17-β-estradiol, and 17-α-estradiol in separate small vials, dissolve in ~0.5 mL of acetone, and cap.

7. Mark a TLC plate (always with a pencil, as ink from a pen will dissolve and run) and spot standard solutions by dipping the spotter in the solution and gently touching the TLC surface. Repeat two to three times for each spot.

8. 15–20 min into the reaction time, carefully remove the condenser and dip the spotter from the crude reaction mixture and spot a designated spot on the TLC plate (should be four spots total). Cap the vial again.

9. Now reaction progress can be monitored by running TLC in the developing solution (acetone:dichloromethane: hexanes – 22:50:28).

10. Develop the TLC plate by taking the plate with forceps and dipping in the developing solution. Then heat with a heat gun until spots appear. If the crude mixture still has a spot corresponding to estrone starting material, repeat the TLC procedure (by now 5–10 min should have passed and

---

* Secure the thermometer with a clamp and place in sand bath or oil bath.

the total reaction time should be nearing the prescribed 30 min). Continue monitoring the reaction until the starting material spot disappears from the crude mixture.

11. Once the starting material is consumed (by TLC analysis), bring reaction to a gentle boil for 10 min.
12. Allow the reaction mixture to cool to room temperature and quench by slow addition of 5 mL of chilled water.
13. Cool the flask in an ice bath for 15 min. Crystals should appear.
14. Collect the crystals by vacuum filtration and rinse with a few milliliters of chilled water.
15. Dry the crystals in the oven at 120°C for 10–15 min. Allow to cool and record their mass.
16. Prepare a small TLC solution of your product by dissolving 4–7 mg in ~0.5 mL of ethanol. Spot a marked TLC plate with your product, the solution of the isomer you think it is (should be evident from the first TLC run), and a co-spot of the two (three spots in total). Run and develop the TLC as before and confirm that your product is predominantly 17-β-estradiol.
17. Note the melting point of your product and comment on its purity.

Characterization[1]:

Estrone: Melting point 260°C–262°C, R_f = 0.61
17-β-estradiol: Melting point 176°C–180°C, R_f = 0.46
17-α-estradiol: Melting point 220°C–223°C, R_f = 0.50*

References

1. Aditya, A.; Nichols, D. E.; Loudon, G. M. *J. Chem. Educ.* **2008**, *85*, 1535–1537.

---

* R_f values are reported for acetone:dichloromethane:hexanes 22:50:28 eluting solution.

Chapter 2

Synthesis of 2,2-Dimethyl-4-pentene-1-amine

Keywords: Air-free techniques, multistep synthesis, multigram synthesis, reduction

Safety: Butyllithium is an extremely strong base and will cause severe burns. Must be handled with extreme care and in a moisture-free environment. May ignite on contact with water. Lithium aluminum hydride is a fine powder and is highly reactive with water (may ignite on contact). Handle with care to avoid making powder airborne and inhaling it. Never weigh more than 2 g at a time. Tetrahydrofuran (THF) is a volatile solvent and is extremely flammable. When using as a solvent for exothermic reactions, make sure to provide adequate cooling. THF is an ether and is prone to forming explosive organic peroxides after prolonged contact with air. Make sure that the THF in use is always fresh. For full MSDS, please refer to www.sigmaaldrich.com.

Equipment Required

- Stirrer hotplate
- 500 and 250 mL round bottomed flasks with stirrer bars
- Large metal or Pyrex bowl for ice bath
- Rubber septa, needles, supply of dry nitrogen or argon gas
- Large filter flask and vacuum for filtration
- Fritted filter
- Large separatory funnel
- Rotatory evaporator
- Litmus paper
- Vacuum pump capable of producing a reduced pressure of at least 1 torr
- NMR spectrometer—optional

Materials Required

- Dry tetrahydrofuran
- Diethyl ether
- Brine
- Dry ice/acetone for cooling
- *n*-BuLi in hexanes
- diisopropylamine (distilled, dry)
- 2-methylpropanenitrile (isobutyronitrile)
- Allyl bromide
- Lithium aluminum hydride
- HCl in ether (best) or dioxane
- Celite®

Purpose: To conduct a multistep synthesis on a large scale and isolate an amine containing a remote carbon–carbon double bonded product; to gain experience of air-free techniques and learn to handle corrosive and pyrophoric materials safely.

Amines containing remote double bonds are widely used in various syntheses of natural products and pharmaceuticals. These types of compounds are extremely versatile and may serve many purposes when equipped with an appropriate protecting group. Thus it is useful to prepare such compounds in gram quantities for further elaboration.

Overall Reaction

The overall procedure includes *in situ* formation of LDA (lithium diisopropylamide) by deprotonating diisopropylamine with *n*-BuLi (butyllithium). LDA acts as a base and deprotonates the nitrile, turning it into a nucleophile. The reaction between the deprotonated nitrile (nucleophile) and allyl bromide (electrophile) results in the formation of a new carbon–carbon bond. This is followed by the reduction of the nitrile group with LAH (lithium aluminum hydride) to produce the desired amine (Scheme 2.1).

The first step involves the use of strong bases, nucleophiles, and electrophiles. In order to be successful, the reaction conditions must be as moisture free as possible. Also, the amounts of base to be used must be measured accurately to avoid side reactions. Furthermore, these reactions are extremely exothermic and it is important to control the reaction temperature (Scheme 2.2).

Next, the nitrile product is reduced to an amine and converted into a nonvolatile salt for storage. This step will allow

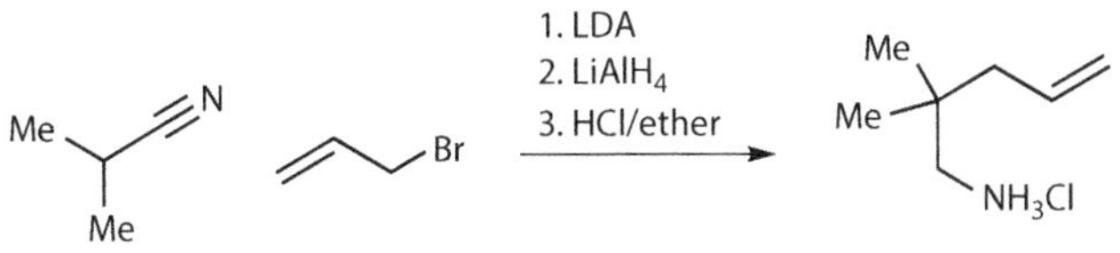

Scheme 2.1 Overall reaction for the synthesis of 2,2-dimethyl-4-pentene-1-amine.

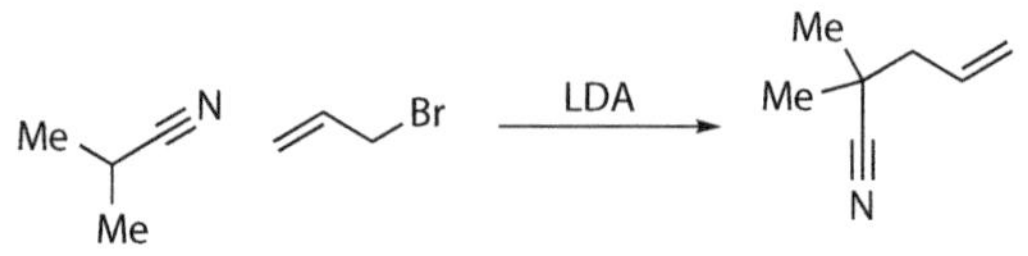

Scheme 2.2   Formation of the intermediate nitrile product.

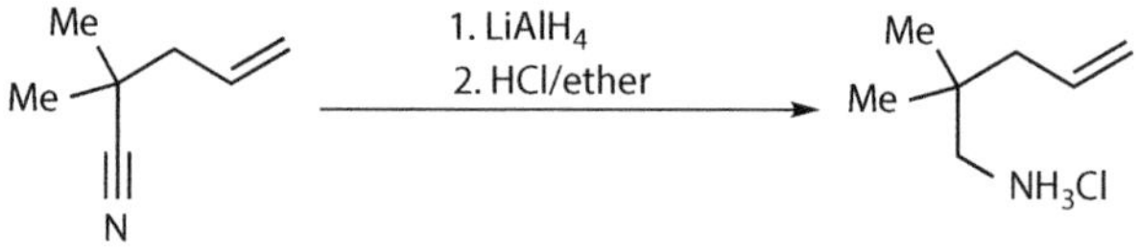

Scheme 2.3   Reduction of the nitrile and subsequent formation of the amine hydrochloride salt.

production of the crude amine that can be further modified with various protecting groups and used in microscale experiments. The reduction of the nitrile proceeds rapidly and easily as LAH is a very powerful reducing agent.

After this step, the product can be converted directly to an appropriately protected form. The amine itself is extremely volatile and we will convert it to its hydrochloride salt by acidification with HCl/ether (Scheme 2.3). The resulting salt is stable and may be stored, but it is extremely hygroscopic and should be stored in an airtight container.

Procedure

Lab Period 1

Caution: *n*-BuLi is an extremely strong base and will cause severe skin damage upon contact. It may also ignite on contact with water. Refer to MSDS for complete safety info.

1. Equip a 500 mL round-bottomed flask with a magnetic stirrer bar, flame-dry it, and allow it to cool to room temperature in a desiccator.

2. Quickly remove the flask from the desiccator and cap it with a rubber septum and clamp it securely to a ring stand.
3. Purge the flask with dry nitrogen or argon gas for a few minutes, then remove the relief valve (usually a needle) while maintaining positive pressure.
4. Charge the flask with 30 mL of dry THF (tetrahydrofuran) and diisopropylamine (21 mmol) and stir. Then cool the mixture to −78°C by placing the flask in a dry ice bath (dry ice and acetone).
5. Slowly add *n*-butyllithium (21 mmol) via syringe. Allow the mixture to react for 30 min.
6. Warm the mixture slowly to 0°C by placing it in an ice bath.
7. Slowly add 2-methylpropanenitrile (20 mmol) in 30 mL of dry THF and allow to react for 1 h at 0°C.
8. Add 2.1 mL of allyl bromide (25 mmol) dropwise and allow it to react for 2 h at 0°C. The mixture may turn a red-brown color. At this point the reaction may be left to warm to room temperature overnight. The mixture can be left until the next lab period.

Lab Period 2

1. To quench the mixture, cool it first to 0°C, open the rubber septum, and slowly add 50 mL of cold deionized water dropwise. In order to diminish the fire hazard, ensure that the reaction mixture stays under nitrogen or argon. White lithium salts will be formed.
2. Pack a large fritted funnel with Celite and set up a large filter flask for vacuum filtration. Filter the lithium salt through the Celite and rinse twice with 50 mL aliquots of ether.
3. Transfer the filtrate into a large separatory funnel and separate the organic layer (top). Wash the aqueous layer with two 50 mL aliquots of ether.

4. Combine the organic layers and transfer them back into the separatory funnel. Wash the organic layer with brine and separate and dry over anhydrous magnesium sulfate or sodium sulfate.
5. Decant or filter the mixture and condense under reduced pressure. Do not apply excessive vacuum as the product is very volatile. It is best to leave some solvent behind rather than lose the product.*

Lab Period 3

Caution: LAH is a very good reducing agent and a base. It will react violently with water! Do not weigh out more than 2 g at a time! LAH is a fine powder and can become airborne very easily. Handle with great care to avoid burning nasal passages and eyes.

1. Purge a dry 500 mL round-bottomed flask equipped with a stirrer bar with dry nitrogen.
2. Weigh out LAH powder (40 mmol) and quickly transfer it into the flask, cap it with a rubber septum and keep it under a dry nitrogen atmosphere. Cool to 0°C in an ice bath.
3. Slowly add 50 mL of dry THF via syringe. Stir gently until all the LAH is suspended. Ensure that there are no LAH clumps!
4. Add 50 mL of dry THF to the crude nitrile mixture and cool it to 0°C in an ice bath. Then add it dropwise to the LAH suspension via syringe. Stir vigorously during this addition. Allow the reaction to proceed for 1 h at 0°C. It may then be allowed to warm to room temperature.

---

* The nitrile product can be stored at low temperatures and securely capped for a later session.

5. To quench the reaction, cool the flask to 0°C in an ice bath, remove the septum, and add 3 M aqueous sodium hydroxide dropwise until the fizzing stops. Make sure to provide adequate cooling and vigorous stirring as the reaction of LAH with water is extremely exothermic and is accompanied by evolution of hydrogen gas. To avoid the risk of fire, keep the mixture under dry nitrogen or argon. White lithium salts will form.

6. When the reaction stops fizzing, filter the material through a large fritted funnel packed with Celite. Rinse twice with 30 mL of ether.

7. Separate the organic layer in a large separatory funnel and wash with 50 mL of brine. Collect the organic layer and dry it over anhydrous magnesium or sodium sulfate.

8. Concentrate the solution to approximately 50 mL under reduced pressure. The product is extremely volatile so make sure not to apply excessive vacuum and temperature. Save a small sample for NMR spectroscopy.

9. Transfer the mixture into a 250 mL round-bottomed flask and cool to 0°C in an ice bath.

10. Stir the mixture and acidify to litmus with dry HCl in ether. A white solid should appear (if the mixture is even slightly wet, the product may be a viscous gum at the bottom of the flask).

11. Once acidified, the product becomes nonvolatile and can be stored for a long time without decomposition or evaporation. To recover the amine salt, concentrate the mixture under reduced pressure and place it under vacuum overnight.

Note: The salt is extremely hygroscopic and may be in a form of a viscous oil if it has been exposed to even the slightest amount of adventitious water. However, it can still be used for further reactions. If a solid is desired, additional rinses with a mixture of ether and hexane and subsequent concentration under reduced pressure have been known to help.

Characterization[1]**:** The spectroscopic data should match the literature values: ^{1}H NMR (300 MHz, $CDCl_3$): δ 5.83–5.74 (m, 1H), 5.05–4.99 (m, 2H), 1.97 (d, J = 7.2 Hz, 2H), 0.94 (br, 2H), 0.85 (s, 6H).

References

1. Tamaru, Y.; Hojo, M.; Higashimura, H.; Yoshida, Z. I. *J. Am. Chem. Soc.* **1988**, *110*, 3994–4002.

Chapter 3

Protection of 2,2-Dimethyl-4-pentene-1-amine

Keywords: Protection of amines, thin layer chromatography, flash column chromatography, ^{1}H-NMR

Safety: *p*-toluenesulfonyl chloride is highly reactive and corrosive. Avoid contact with skin by wearing gloves. It will cause severe eye damage upon contact. For full MSDS, please visit www.sigmaaldrich.com.

Equipment Required

- Stirrer hotplate
- 50 and 250 mL round bottomed flasks with stirrer bars
- 50 mL beaker
- Medium metal or Pyrex bowl for ice bath
- Rubber septa, needles, plastic syringes, supply of dry nitrogen or argon gas
- Large filter flask and vacuum for filtration

- 125 mL separatory funnel
- Medium chromatography column
- Solvent bulb
- One hundred 10 mL test tubes and a rack
- Pasteur pipettes
- Rotatory evaporator
- Vacuum pump capable of producing a reduced pressure of at least 1 torr
- Supply of positive air pressure
- Rubber hose and ground glass adapter to fit the column
- Heat gun
- NMR spectrometer

Materials Required

- Crude 2,2-dimethyl-4-pentene-1-amine HCl salt from experiment 2
- *p*-toluenesulfonyl chloride
- Dry dichloromethane
- Freshly distilled, dry, triethylamine (TEA) or diisopropyle-thylamine (DEA); see Appendix
- Saturated aqueous sodium bicarbonate solution
- Ice for cooling
- Silica (60 mesh)
- Sand
- TLC plates
- $KMnO_4$ solution for developing TLC/UV lamp
- Ethyl acetate
- Hexanes (mixture of isomers)
- Anhydrous $MgSO_4$ or Na_2SO_4
- 1 M HCl

Purpose: To protect the free amine group so it does not interfere with further functionalization of the molecule.

Protection of functional groups is a very common technique used in organic synthesis. It is usually referred to as a "necessary evil" because protection, and subsequent deprotection, of a functional group adds two steps to the total synthesis. However, it is necessary to mask functional groups if they interfere with reactions in other parts of the molecule. There is a plethora of protecting groups for virtually every functional group. For a good reference material, look up *Greene's Protective Groups in Organic Synthesis* published by Wiley.

Overall Reaction

Since 2,2-dimethyl-4-pentene-1-amine hydrochloride salt is used, at least two equivalents of base (TEA) are required. The first equivalent deprotonates the salt releasing the free amine, which is a reasonable nucleophile. The second equivalent is necessary to "mop up" hydrochloric acid formed when the amine attacks TsCl (track the protons and balance the equation in Scheme 3.1).

p-toluenesulfonyl chloride is an excellent electrophile. The sulfur atom is highly electron poor and chloride is a good leaving group. Even a modest nucleophile, such as a free amine, can displace the chloride in very good yields under

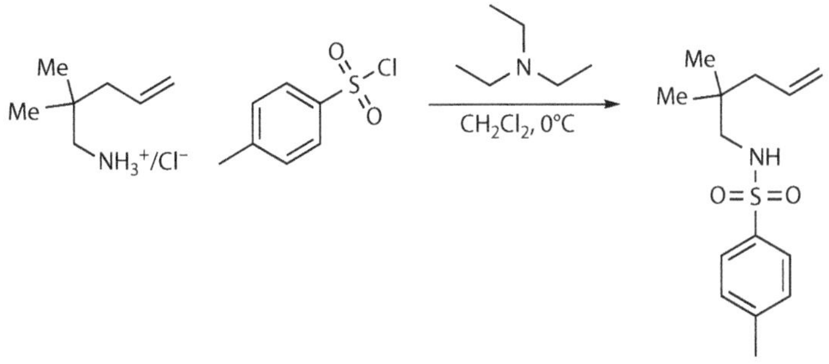

Scheme 3.1   Formation of a toluenesulfonamide protecting group.

mildly basic conditions. Tosylation of nitrogen makes it a poor nucleophile (and Lewis base), which allows subsequent reactions intolerant of free amines (such as the use of transition metal catalysts).

Procedure

Lab Period 1: Protection of 2,2-dimethyl-4-pentene-1-amine

1. Equip a flame-dried and cooled 50 mL round-bottomed flask with a stirrer bar, cap it with a septum and purge with nitrogen or argon and keep under positive pressure. Add 10 mL of dry dichloromethane (DCM) via syringe.
2. Weigh out 0.75 g of 2,2-dimethyl-4-pentene-1-amine HCl salt (0.5 mmol) and transfer into the flask. Cap the flask with the septum. Cool to 0°C in an ice bath.
3. With stirring provided, slowly add 1.8 mL of freshly distilled, dry triethylamine (1.3 mmol) via syringe.
4. Weigh out 0.96 g of *p*-toluenesulfonyl chloride (0.5 mmol) in a small beaker. Dissolve it in 5 mL of dry DCM and quickly draw the solution into a dry syringe.
5. Add the *p*-toluenesulfonyl chloride solution dropwise into the flask through the septum. Rinse the flask with an additional 3 mL of DCM and add it to the flask.
6. Remove the flask from the ice bath, allow the reaction mixture to warm up to ambient temperature, and allow to it stir overnight (or until the next lab session).

Lab Period 2: Isolation and Purification of N-(2,2-dimethylpent-4-enyl)-2-toluenesulfonamide

1. After at least 4 h reaction time, cool the mixture to 0°C in an ice bath and quench with 10 mL of 1 M hydrochloric acid.

2. Transfer the mixture into a separatory funnel and isolate the organic layer (DCM is on the bottom). Place the organic layer back into the separatory funnel and wash with an additional 10 mL of 1 M HCl.
3. Separate and dry the organic layer over anhydrous $MgSO_4$ or Na_2SO_4.
4. Decant or filter the mixture into a 250 mL round-bottomed flask and concentrate to dryness under reduced pressure. The crude mixture may either crystallize or remain as a thick oil.
5. Prepare a 1 mL solution of TsCl in DCM and spot a TLC plate with it. Make another spot with the crude mixture a few millimeters away on the same plate. Run the plate in 15% (*v*) ethyl acetate/hexanes. Develop in $KMnO_4$ solution with subsequent heating over a heat gun or under UV light. The product and TsCl should be half way up the plate, while the TEA salt and unreacted starting material (should there be any) will stick to the baseline.
6. Pack the column with a slurry of silica in 10% ethyl acetate/hexane. Make sure *never* to let the column run dry. See the "Appendix: Essential Laboratory Techniques" for full procedure.
7. Dissolve the crude reaction mixture in a minimum amount of DCM and load the column. Run the column with 10% ethyl acetate/hexanes. Monitor the fractions by TLC. Combine the fractions containing the product in a 1 L round-bottomed flask and concentrate under reduced pressure.
8. Dissolve the solid product in a minimum amount of DCM and transfer into a vial. Concentrate under reduced pressure. Then place under vacuum for at least 3–4 h. Transfer a small crystal of the purified and dried product into an NMR tube, add $CDCl_3$, and record the ^{1}H-NMR of the product to confirm identity and purity.

Characterization: The product is a white solid.[1]

^{1}H NMR (300 MHz, CDCl$_3$): δ 7.74 (d, J = 8.4 Hz, 2H), 7.33 (d, J = 8.1 Hz, 2H), 5.72 (ddt, J = 10.2, 17.7, 7.2 Hz, 1H), 5.04–4.97 (m, 2H) 4.41 (br, 1H), 2.68 (d, J = 6.9 Hz, 2H), 2.43 (s, 3H), 1.95 (d, J = 7.2 Hz, 2H), 0.86 (s, 6H).

References

1. Michael, F. E.; Cochran, B. M. *J. Am. Chem. Soc.* **2006**, *128*, 4246–4247.

Chapter 4

Optimization of the Reaction Conditions for the Aminofluorination of *N*-(2,2-Dimethylpent-4-enyl)-2-toluenesulfonamide

Keywords: Aminofluorination, thin layer chromatography, flash column chromatography, ^{1}H-NMR spectroscopy, microscale synthesis, optimization

Safety: Tetrafluoroboric acid is a very strong acid and is extremely corrosive. Furthermore, it forms HF on contact with water, which is extremely dangerous. Wear protective clothing and gloves. Have sodium gluconate on hand in case of exposure.

Silver fluoride is corrosive and readily absorbs water from atmosphere.

Hypervalent iodine compounds, although benign, are strong oxidants and may be explosive (depending on type and conditions of storage). Store these compounds in tightly sealed containers at 4°C–6°C.

For full MSDS, please visit www.sigma.com.

Equipment Required

- Stirrer hotplate
- 4 × 10 mL and 50 mL round-bottomed flasks with stirrer bars
- Medium metal or Pyrex bowl for ice bath
- Rubber septa, needles, plastic syringes, supply of dry nitrogen or argon gas
- Pasteur pipettes
- Glass funnel
- 10 and 50 µL Hamilton syringes
- Rotatory evaporator
- Vacuum pump capable of producing a reduced pressure of at least 1 torr
- Heat gun
- NMR spectrometer

Materials Required

- *N*-(2,2-dimethylpent-4-enyl)-2-toluenesulfonamide from experiment 3
- Tetrafluorboric acid diethyl ether complex
- Iodosyl benzene (will need to be synthesized at least 2 days prior)[3]
- (diacetoxyiodo)benzene
- [bis(trifluoroacetoxy)iodo]benzene (optional)
- bis(*tert*-butylcarbonyloxy)iodobenzene
- Dry dichloromethane
- Dry acetonitrile
- Silver fluoride

- Palladium (II) diacetate
- 1,3-dinitrobenzene
- Saturated aqueous sodium bicarbonate solution
- Ice for cooling
- TLC plates
- $KMnO_4$ solution for developing TLC
- Kim wipes

Purpose: To find optimal conditions for the aminofluorination of *N*-(2,2-dimethylpent-4-enyl)-2-toluenesulfonamide and to compare this with to a transition metal-catalyzed method.

Background: Usually, there are many synthetic paths toward the same molecule. There are many factors to consider when choosing a synthetic strategy. Here are just a few: cost of starting materials and reagents, number of steps, stereo- or regioselectivity, overall yield, and yield for each individual step, safety, and waste toxicity.

Scientists are constantly working on discovering new methods and perfecting existing ones. However, even if an idea is shown to be promising, optimum conditions need to be found. The process of optimization can be long and tedious but when you pool the results from the whole lab group, the best conditions can be found more quickly (Scheme 4.1).

Until 2009, there was no direct way to synthesize 3-fluoropiperidines from aminoalkenes in a single step. Liu and coworkers developed a palladium-catalyzed method that gave a wide range of sulfonamide-protected fluoropiperidines in respectable yields.[1] Silver fluoride was used as the source of "F⁻", hypervalent iodine as the oxidant, and palladium (II) as the catalyst (Scheme 4.2).

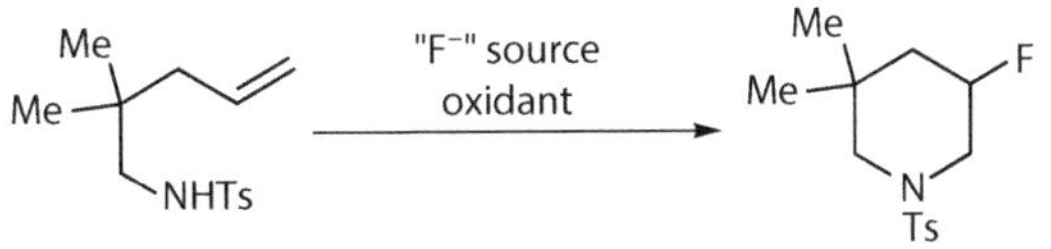

Scheme 4.1 General reaction for aminofluorination.

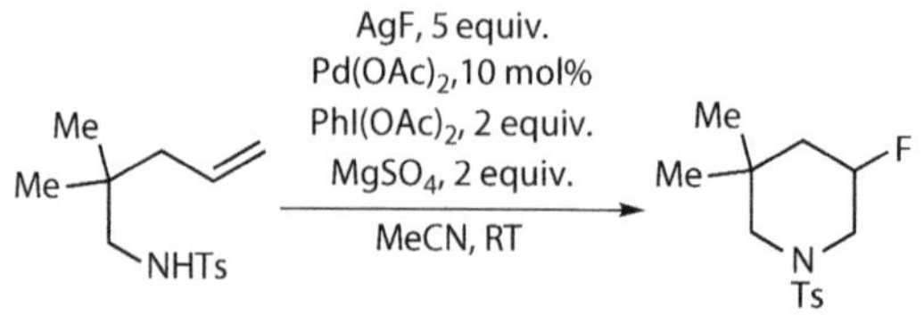

Scheme 4.2 Palladium-catalyzed aminofluorination.

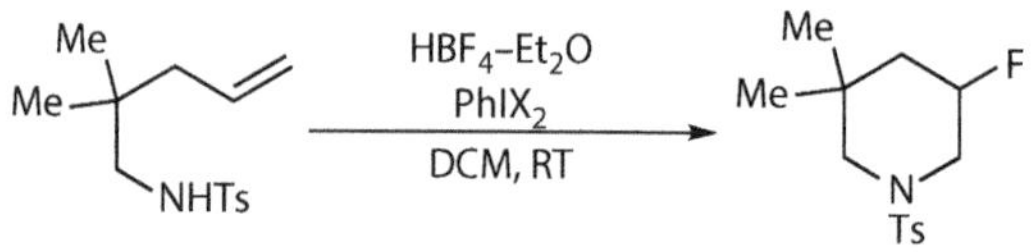

Scheme 4.3 Metal-free aminofluorination.

You suspect that other "F⁻" sources would be suitable for this method. Now you and your lab mates will perform a series of reactions that will determine optimal conditions for the new method (Scheme 4.3). In order to expedite the process, you will be using ^{1}H-NMR spectroscopy with an internal standard to determine yields. Once optimal conditions are found, an isolated yield will be obtained. In order to validate your lab skills and compare the "old" method to the "new" one, the original results need to be replicated.

Procedure

Lab Period 1: Part 1a Steps 1–4 and Part 1b
Lab Period 2: Part 1a Steps 5–11

Part 1a: Aminofluorination of N-(2,2-dimethylpent-4-enyl)-2-toluenesulfonamide: Palladium-Catalyzed Method

1. Equip a flame-dried and cooled 10 mL round-bottomed flask with a stirrer bar, cap it with a septum and purge

with nitrogen or argon, and keep under positive pressure of the dry gas.

2. Weigh out
   - 24.7 mg of *N*-(2,2-dimethylpent-4-enyl)-2-toluenesulfonamide (0.1 mmol)
   - 81.2 mg of bis(*tert*-butylcarbonyloxy)iodobenzene (0.2 mmol)
   - 2.3 mg of palladium (II) acetate (0.01 mmol, 10 mol%)
   - 24.0 mg of anhydrous magnesium sulfate (0.2 mmol)
   - 16.8 mg of 1,3-dinitrobenzene (0.1 mmol) and transfer it to the flask.

3. Weigh 63.5 mg of silver fluoride (0.5 mmol) quickly (it is hygroscopic and will stick to weighing paper) and place it in the flask.

4. Add 0.5 mL of dry acetonitrile and stir the reaction for at least 24 h at room temperature.

5. After 24 h, subject the crude reaction to TLC in 15% (*v*) ethyl acetate/hexanes to verify that all starting material is gone.

6. After ensuring that the reaction has gone to completion, clamp a 50 mL round-bottomed flask. Loosely pack a glass funnel with Kim wipe and place it in the funnel.

7. Filter the reaction mixture. Rinse the flask three times with 2–3 mL of dichloromethane and run them through the filter to wash the solid residue.

8. Concentrate the filtrate under reduced pressure and place the crude mixture under vacuum for at least 1 h.

9. Dissolve the entire sample in deuterated chloroform, transfer into an NMR tube and record the ^{1}H-NMR spectrum of the crude mixture. *Make sure to run it only on 1 scan* (the 1,3-dinitrobenzene singlet that you will be using for integration and determination of NMR yield tends to under-integrate when multiple scans are performed).

10. Integrate the spectrum and assign the spectrum. Try to identify any impurities.

- 25 and 50 mL Erlenmeyer flasks
- 125 mL filter flask
- Rotatory evaporator
- UV lamp
- HPLC instrument with a UV detector and Chiralcel® OD-H column or similar
- NMR spectrometer, NMR tube and CDCl$_3$

Materials Required

- (*S*)-Proline
- 2-methyl-1,3-cyclohexadione
- 3-Buten-2-one
- Wieland–Miescher ketone (a pure enantiomer if possible)
- Acetonitrile
- Diethyl ether
- Ammonium chloride, saturated aqueous solution
- Silica
- Sand
- TLC plates
- UV lamp
- Ethyl acetate
- Hexanes (mixture of isomers)
- Hexanes (HPLC grade)
- Isopropanol (HPLC grade)
- Anhydrous MgSO$_4$ or Na$_2$SO$_4$

Purpose: To use a chiral catalyst in the synthesis of the Wieland–Miescher ketone via a one-pot Robinson annulation and to determine the enantiopurity of the product by HPLC.

Overall Reaction

The Wieland–Miescher ketone (Figure 5.1) is a bicyclic enedione with one chiral center. The structure contains the

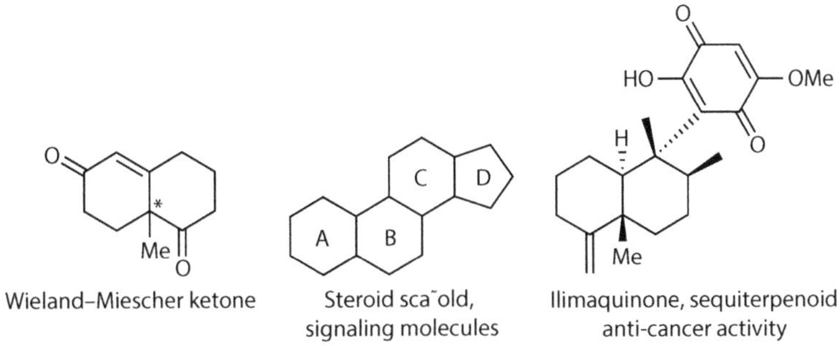

Figure 5.1 Wieland–Miescher ketone.

A/B ring substructure of the steroidal carbon skeleton. This compound has been used in the total synthesis of steroids and terpenoids.[1] Thus, an enantioselective route that is facile and effective makes it an attractive synthon for industrial-scale reactions (Scheme 5.1).

The Wieland–Miescher ketone can be synthesized using a Robinson annulation of simple starting materials. Enantioselectivity comes from the use of a chiral catalyst. This lab will employ a simple one-pot reaction with (*S*)-proline as the catalyst.

The Robinson annulation is a reaction between a ketone and a methyl vinyl ketone. It was developed by Robert Robinson in the first part of the twentieth century and is an efficient route to α,β-unsaturated ketones in six-membered rings. The reaction proceeds in two steps: Michael addition (conjugated addition of enolate to α,β-unsaturated ketone, methyl vinyl ketone in this case) followed by aldol condensation and dehydration.

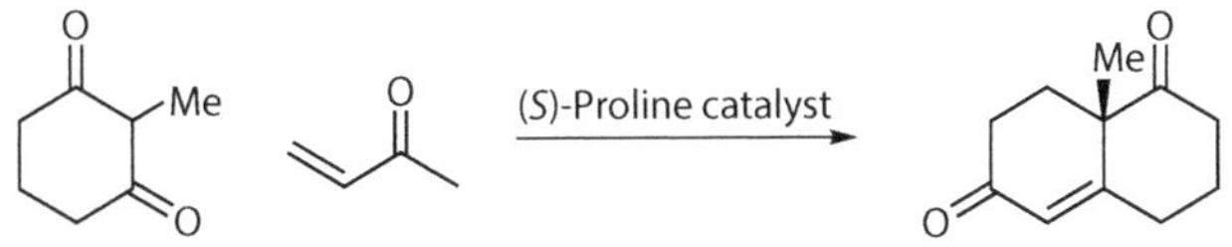

Scheme 5.1 Synthesis of the Wieland–Miescher ketone by Robinson annulation.

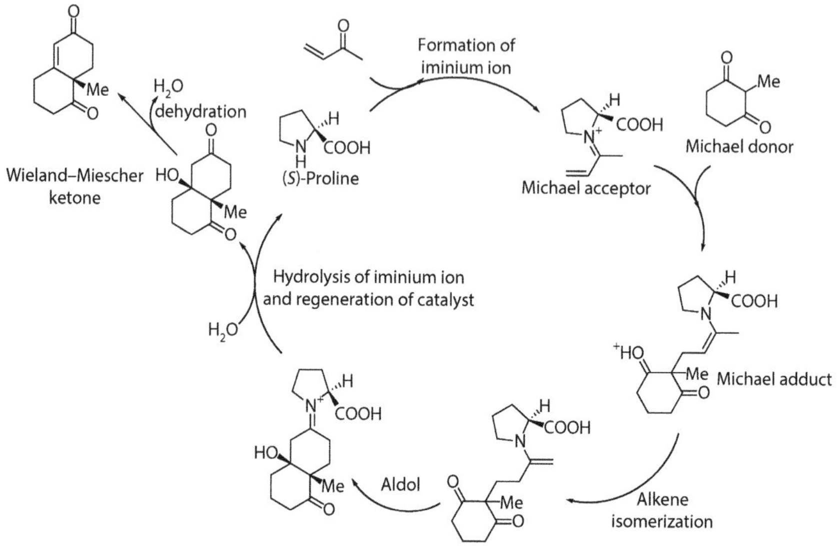

Scheme 5.2   Mechanism of proline-catalyzed Robinson annulation. (From Lazarski, K.E. et al., *J. Chem. Educ.*, 85, 1531, 2008.)

The proline-catalyzed annulation is slightly different from standard base-catalyzed reaction. The proline catalyst forms iminium ion by reacting with the carbonyl of the Michael acceptor. Since proline is chiral, it will induce enantioselectivity in the cyclization (aldol) step. Carefully review the proposed mechanism above and try to supplement it with electron-pushing arrows (Scheme 5.2).

Procedure

Lab Period 1

1. Set up a hot water bath in a 50 mL beaker and maintain it at 35°C.
2. Equip a 25 mL round-bottomed flask with a stirrer bar and weigh into it 40.3 mg (0.35 mmol, 35 mol%) of (*S*)-proline. Add 5 mL of acetonitrile. Stir until a uniform suspension is formed.

3. Weigh 126.2 mg (1 mmol) of 2-methyl-1,3-cyclohexadione and add to the stirring mixture. Seal the flask with a septum, place in the water bath, and stir for 30 min.
4. Draw 125 μL (1.5 mmol) of 3-buten-2-one and add it to the stirred mixture. Stir at 35°C until the next lab period (at least 48 h).

Lab Period 2

1. Prepare a Pasteur pipette silica column and clamp it firmly.
2. Add 1 mL of saturated $NH_4Cl_{(aq)}$ to the reaction mixture and stir for a few minutes.
3. Transfer the mixture into a separatory funnel. Wash the flask with ether 3 mL × 5 mL and combine the layers in the funnel. Add 5 mL of saturated $NH_4Cl_{(aq)}$ to the funnel and shake a few times with venting.
4. Separate the layers. Extract the aqueous layer with 10 mL of ether and combine the organic layers.
5. Transfer the organic layers back into the funnel and wash with 10 mL of brine. Separate and dry the ether layer with anhydrous magnesium or sodium sulfate.
6. Carefully decant or filter the liquid into a dry 100 mL round-bottomed flask and concentrate under reduced pressure (rotatory evaporator).
7. Prepare 15 mL of a 60/40 mixture of ethyl acetate/ hexanes and dissolve the crude residue in the minimal amount of this mixture (about 0.5 mL).
8. Add the solvent mixture to the Pasteur pipette column and wait until all the silica is wet (several additions may be necessary). You may gently force the solvent by loosely placing a hose with very gentle stream of air over the pipette.
9. Once the solvent layer is almost at the silica line, load the solution of crude reaction mixture, wash the reaction flask with an additional 0.5 mL of eluting solution, and add to the column when the solvent level reaches the silica.

10. Gently add eluting solution so as not to disturb the silica surface and force the liquid through the column under gentle pressure. Run the column until all the eluting solution is used. Collect all the solvent into a 50 mL round-bottomed flask and evaporate to dryness under reduced pressure.
11. Run a TLC plate against a pure sample of Wieland–Miescher ketone and visualize under UV light. Use a 60/40 mixture of ethyl acetate/hexanes.
12. Prepare an NMR sample in $CDCl_3$. Compare the ^{1}H-NMR spectrum of your product with the established literature data.
13. Prepare an HPLC sample by dissolving a few milligrams of the product in 1 mL of ethyl acetate in an appropriate vial.
14. Protocol used for HPLC: Chiralcel OD-H column at 1 mL/min with 5% isopropanol in hexanes. The detector was set to 225 nm wavelength. The enantiomeric excess can be calculated from the area of the peaks.[1]
15. To determine which enantiomer is the major one, spike the sample with a few milligrams of pure enantiomer (if available) of Wieland–Miescher ketone. Observe the growth of one of the peaks. These data will establish which enantiomer is more abundant.

Characterization: (*S*)-(+)-8a-Methyl-3,4,8,8a-tetrahydro-1,6(2*H*,7*H*)-naphthalenedione[1]

^{1}H NMR (300 MHz, $CDCl_3$): δ 5.88 (s, 1H), 2.7–2.6 (m, 2H), 2.55–2.3 (m, 4H), 2.2–2.0 (m, 2H), 1.8–1.6 (m, 2H), 1.4 (s, 3H).

References

1. Kim, M.; Kawada, K.; Gross, R. S.; Watt, D. S. *J. Org. Chem.* **1990**, 55, 504–511.
2. Lazarski, K. E.; Rich, A. A.; Mascarenhas, C. M. *J. Chem. Educ.* **2008**, 85, 1531–1534.

Chapter 6

Formation of *N*-Benzylcinchonidinium Chloride from Cinchonidine

Keywords: Protection of amines, thin layer chromatography, flash column chromatography, ^{1}H-NMR spectroscopy

Safety: Cinchona alkaloids are acutely toxic. Benzyl chloride is a powerful lachrymator, is acutely toxic, and will cause skin irritation. *N*-benzylcinchonidinium chloride should be handled with care as it may cause skin and eye irritation. Acetone is a volatile, flammable solvent and will cause serious eye irritation. Acetone may also cause drowsiness or dizziness. For full MSDS, please visit www.sigmaaldrich.com.

Equipment Required

- Stirrer hotplate
- 250 mL round-bottomed flask with stirrer bar
- Pasteur pipettes

- Reflux condenser
- Drying tube
- Large filter flask and vacuum for filtration
- Büchner funnel and filter paper
- Oil bath
- Thermometer
- Vacuum pump capable of producing at least 1 torr of reduced pressure
- NMR spectrometer
- Polarimeter

Materials Required

- Dry acetone
- Benzyl chloride
- (–)-cinchonidine

Purpose: To synthesize an enantiopure salt on a gram scale to be used in further chiral resolution.

Cinchona alkaloids are chiral compounds that are widely used in asymmetric syntheses and chiral resolutions. This popularity of cinchona alkaloids can be attributed to their abundance and affordability.

Very often one of these alkaloids, cinchonidine, is modified as its chloride salt, which in turn is used in chiral resolution of BINOL, another popular organocatalyst (next experiment). Enantiopure N-benzylcinchonidinium chloride costs around $12/g, while a simple reaction of cinchonidine ($3/g) with benzyl chloride will yield enantiopure N-benzylcinchonidinium chloride (Scheme 6.1).

This is a very simple reaction and can easily be set up in combination with another lab experiment. Working on different reactions at the same time is yet another valuable skill to have in an organic lab. *Please note that the reaction requires at least 48 h.*

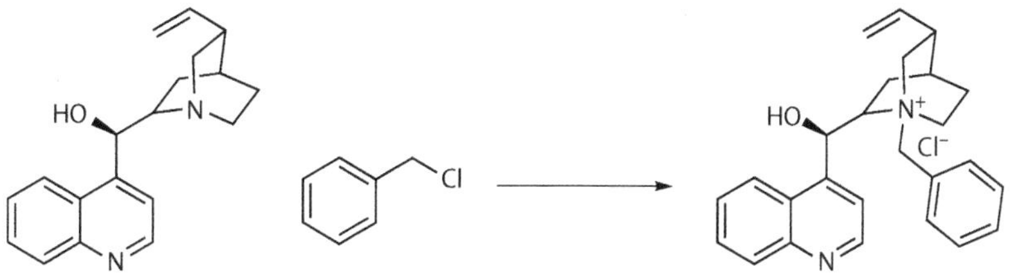

Scheme 6.1   Formation of *N*-benzylcinchonidinium chloride.

Overall Reaction*

Procedure

Lab Period 1

1. Charge a 250 mL round-bottomed flask with a stirrer bar and directly weigh 1.90 g (15 mmol) of benzyl chloride into it, carefully adding it with a Pasteur pipette. This technique will minimize transfer losses. Be careful not to spill the benzyl chloride.
2. Add 2.94 g (10 mmol) of cinchonidine and 70 mL of dry acetone to the flask.
3. Fit the flask with a reflux condenser and a drying tube.
4. Place the oil bath on the hot plate and immerse the thermometer. Place the flask in the bath, turn on the stirring, and gently heat until reflux is reached. The boiling point of acetone is 56°C.
5. Heat under reflux for at least 2 days. It is advisable to check on the reaction at least twice a day to ensure that all is normal. Follow local safety rules concerning reactions left unattended overnight.

Lab Period 2

1. Turn off heat and allow the reaction to cool to room temperature.

---

* *Note*: Reaction setup can take place in combination with another lab experiment.

2. Set up a Büchner funnel and collect the precipitate by filtration.
3. Wash the precipitate twice with 20 mL portions of dry acetone and allow to air dry for a few minutes.
4. Transfer the product into a vial or flask and place under vacuum for at least 30 min.
5. Confirm the purity of the salt by ^{1}H NMR spectroscopy (in d_6-DMSO) and its optical rotation by polarimetry.
6. If you choose to perform resolution of BINOL (next experiment), this is a good time to set up.

Characterization: The product is an off-white solid. Compare the ^{1}H NMR spectrum to the one found at www.sigma.com.

According to www.sigma.com, the optical rotation of (8*S*,9*R*)-(−)-*N*-benzylcinchonidinium chloride is α_{21}^D −180°, $c = 1.3$ in H_2O.

Chapter 7

Chiral Resolution of BINOL (1,1′-bi-2-Naphthol) with *N*-Benzylcinchonidinium Chloride

Keywords: Chiral resolution, polarimeter, optical activity, ^{1}H-NMR spectroscopy, gram scale

Safety: BINOL has been reported to be acutely toxic. *N*-benzylcinchonidinium chloride should be handled with care as it may cause skin and eye irritation. Acetonitrile and ethyl acetate are flammable, toxic solvents and will cause severe eye irritation. For full MSDS, please visit www.sigmaaldrich.com.

Equipment Required

- Stirrer hotplate
- 100 mL round bottom flasks

- 125–250 mL Erlenmeyer flask
- Oil bath
- Thermometer
- Büchner funnel and filter flask
- 125–250 mL separatory funnel
- Stirrer bar
- Reflux condenser
- Rotatory evaporator
- Large metal bowl or Pyrex dish for ice bath
- NMR spectrometer
- Polarimeter
- Source of vacuum

Materials Required

- Racemic BINOL
- *N*-benzylcinchonidinium chloride (from previous experiment or purchased)
- Acetonitrile
- Ethyl acetate
- Brine
- Sodium or magnesium sulfate, anhydrous
- 1 M hydrochloric acid

Purpose: To resolve a racemic mixture of 1,1′-bi-2-naphthol (BINOL) into its enantiomers using cinchona alkaloids.

Use of chiral auxiliaries and catalysts to induce enantioselectivity and diastereoselectivity has become very popular in the past few decades. BINOL is one of those compounds.[1] The challenge, however, is to complete a project on time and on budget: (*R*)-BINOL currently costs $54/g and (*S*)-BINOL— $46.5/g, while a mixture of enantiomers costs around $5/g (at the time of publication in the United States). Hence, chiral resolution becomes a reasonable option for obtaining enantiopure organocatalysts.

It is interesting to note that the chiral nature of BINOL is not exactly obvious: there are no sp^3-hybridized carbons, thus no chiral centers. Despite that, the molecule has no plane of symmetry and is optically active. The rotation around the biaryl bond is restricted due to steric hindrance; hence the two conformational isomers cannot interconvert (atropisomerism).

Overall Reaction

The resolution of a racemic mixture of BINOL involves the formation of separable diastereomers (Scheme 7.1). Recall that enantiomers possess exactly the same chemical and physical properties and hence are impossible to separate based on solubility, reactivity, boiling point, etc. However, diastereomers are different compounds with different chemical and physical properties. In this case, we will separate the diastereomers taking advantage of a difference in their solubility.

A racemic mixture of BINOL will be reacted with enantiopure *N*-benzylcinchonidinium chloride to form diastereomeric salts. One diastereomer has a lower solubility than the other complex and can be isolated by crystallization.

Once the single diastereomeric salt is collected, it will need to be hydrolyzed to free enantiopure BINOL. The final product will be analyzed by NMR spectroscopy for any impurities and determination of the optical rotation will show whether or not it is enantiopure.

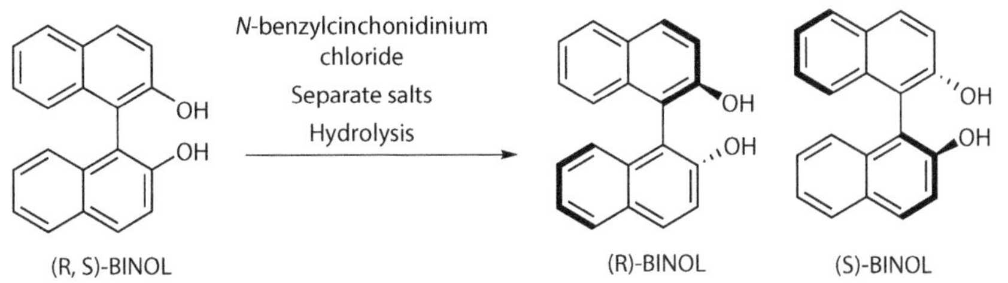

Scheme 7.1   Resolution of BINOL.

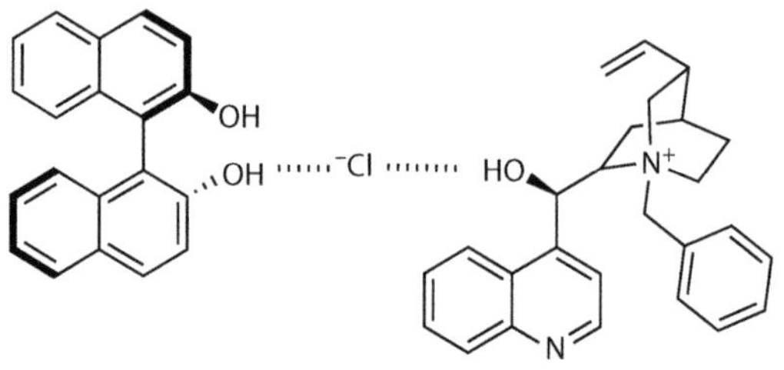

Figure 7.1 *(R)*-BINOL–*N*-benzylcinchonidinium complex.

Cinchona alkaloids, cinchonine and cinchonidine, are pseudoenantiomers (not exact mirror images due to steric inhibition of rotation) and have been historically used to resolve BINOL derivatives. However, it is also interesting to note that when used as *N*-benzyl chloride salt, both cinchonidine and cinchonine form a crystalline salt only with *(R)*-BINOL (Figure 7.1).[2]

Procedure*

Lab Period 1

1. Equip a 100 mL round-bottomed flask with a stirrer bar and reflux condenser; add 2.9 g (10 mmol) of racemic 1,1'-bi-2-naphthol (BINOL) and 2.4 g (5.5 mmol) of *N*-benzylcinchonidinium chloride.
2. Add 40 mL of acetonitrile and heat the mixture under reflux with stirring for 4 h (acetonitrile bp is 81–82°C for reference). In order to heat the mixture under reflux, it is easiest to set up an oil bath on the stirrer hot plate, equip it with a thermometer, and monitor the temperature until reflux is reached and stabilized.

---

* *Note*: The procedures for lab period 1 could be easily combined with another lab due to long waiting periods and short setup time. If using *N*-benzylcinchonidinium chloride from Experiment 4, the setup can be started at the end of that lab.

3. After the suspension is completely dissolved (about 4 h), cool to ambient temperature and stir overnight (or until the next lab period).

Lab Period 2

1. Cool the mixture to 0°C in an ice bath for 2 h.
2. Set up a Büchner funnel and filter the mixture. This should separate the (*R*)-salt (solid). (*S*)-salt will remain in solution and will not be isolated.
3. Concentrate the solid (*R*)-salt filtrate under reduced pressure, redissolve in 40 mL of ethyl acetate, and transfer into a separatory funnel.
4. Wash the solution twice with 15 mL aliquots of 1 M HCl and 10 mL of brine. Acid wash will hydrolyze the salt and remove the N-benzylcinchonidinium leaving solid (*R*)-BINOL product behind.
5. Transfer the solution into an Erlenmeyer flask and dry over Na_2SO_4 or $MgSO_4$. Filter or carefully decant into the collection flask and concentrate under reduced pressure.
6. Check the purity of the product by ^{1}H-NMR spectroscopy and its optical purity by polarimetry.

Characterization: The product is a light brown solid.[3]

^{1}H NMR (250 MHz, $CDCl_3$) δ: 5.0 (s, OH, 2H), 7.16 (d, *J*=8.3, 2H), 7.30 (m, 2H), 7.38 (m, 4H), 7.90 (d, *J* = 8.1, 2H), 7.99 (d, *J*=8.9, 2H).

Optical rotation according to www.sigma.com: α_{21}^D +34°, $c=1$ in THF.

References

1. Brunel, J. M. *Chem. Rev.* **2005**, 105, 857–898.
2. Hughes, D. L. *Org. Synth.* **2014**, 91, 1–11.
3. Cai, D.; Hughes, D. L.; Verhoeven, T. R.; Reider, P. J. *Org. Synth.* **1999**, 76, 1–3.

Chapter 8

Transesterification of Phosphatidylcholine and FAME Quantification

Keywords: Microscale, biomolecule, transesterification, quantitative analysis, internal standard, GC-MS

Safety: Eggs could contain *Salmonella*. Sulfuric acid is very corrosive and will cause severe burns. Hexanes are highly flammable and must be used in the hood. For full MSDS, please visit www.sigmaaldrich.com.

Equipment Required

- Analytical balance
- Stirrer hotplate
- 100 mL beaker
- Four dram vial with a Teflon-lined screw cap
- Stirrer bar
- 15 mL centrifuge tubes
- Pasteur pipettes

- Gas chromatograph with an FID and mass spectrometer
- FAMEWAX column or equivalent
- Centrifuge, capable of spinning 10 mL samples

Materials Required

- 2% (v/v) solution of H_2SO_4 in methanol
- Saturated aqueous solution of $NaHCO_3$
- One egg
- Hexanes
- 9-fluorenone
- Hexadecanoic (palmitic) acid methyl ester
- Octadecanoic (stearic) acid methyl ester
- 9-(Z)-Octadecenoic (oleic) acid methyl ester
- 9,12-(Z,Z)-Octadecadienoic (linoleic) acid methyl ester

GC method used: Starting temperature 195°C, ramped to 240°C over 6 min and held for 9 min with total experiment time of 15 min. Injector and detector temperature— 240°C. Helium carrier gas at 40 cm/s. Column used: 0.25 μm FAMEWAX, 30 m, 0.25 mm ID. Column was purchased from www.restek.com

Purpose: To transesterify phosphatidylcholine from egg yolk and assay its fatty acid composition by measuring fatty acid methyl esters (FAMEs).

Overall Reaction

Performing reactions on the microscale and working with biomolecules is an increasingly important skill for organic chemists, due to the growing overlap between organic chemistry and chemical biology. Furthermore, it is important to be able

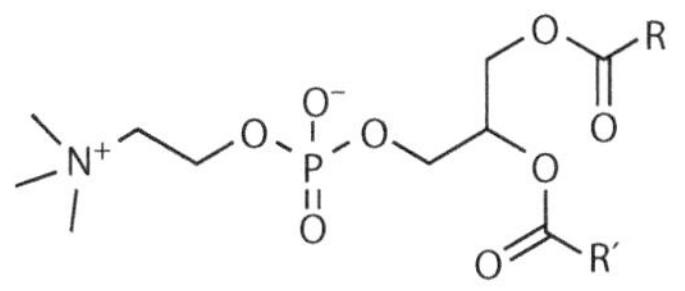

Figure 8.1 Phosphatidylcholine.

to identify isolated compounds and devise methods for their quantitative analysis.

In this experiment, the fatty acid composition of egg yolk will be assayed by GC-MS. Egg yolk is rich in fats, which are mostly in form of phospholipids (phospatidylcholine, Figure 8.1).

Each molecule of phospatidylcholine contains two fatty acid side chains joined through ester linkages. The most abundant fatty acids in egg yolk are as follows (approximate composition): palmitic—26%, stearic—9%, oleic—44%, and linoleic—11%[1] (Figure 8.2). These esters will be transesterified under acidic conditions with methanol to give FAMEs (Scheme 8.1). These methyl esters can then be analyzed by GC-MS.

To quantify these FAMEs, an internal standard is needed. 9-fluorenone (Figure 8.3) is a good fit for this experiment as it is unreactive, does not interfere with FAME peaks, and is readily available for a fraction of the cost of analytical grade standards.

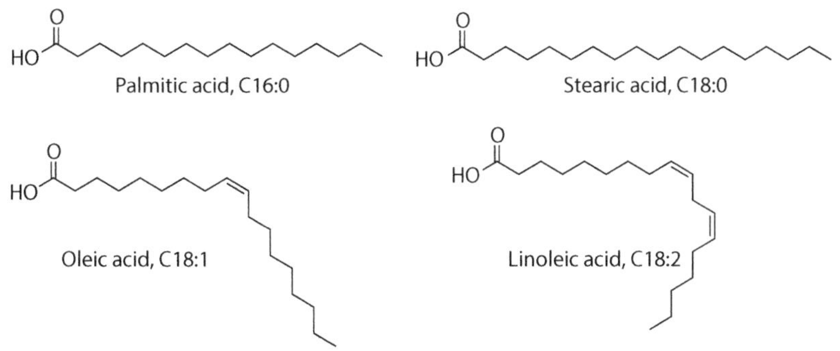

Palmitic acid, C16:0

Stearic acid, C18:0

Oleic acid, C18:1

Linoleic acid, C18:2

Figure 8.2 Fatty acids comprising most of the fat content of egg yolk.

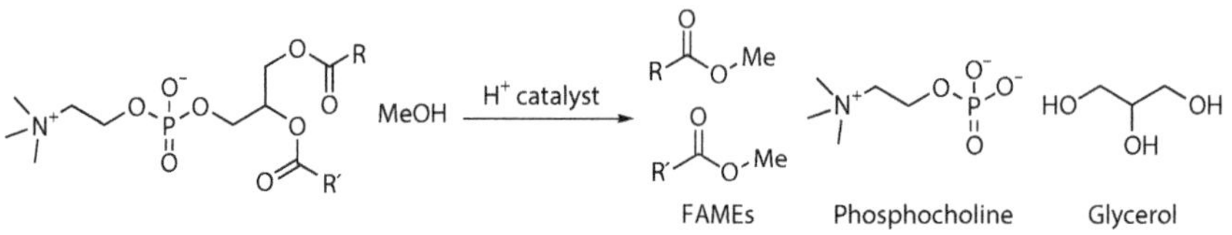

Scheme 8.1 **Transesterification of phosphatidylcholine with methanol.**

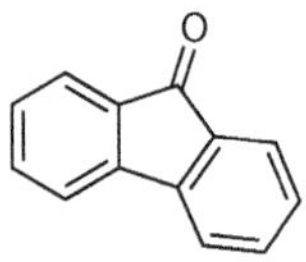

Figure 8.3 **9-fluorenone.**

While the detector response in GC should be linear and based on molar concentration of analytes, this is not always the case. Hence, a standard curve will be needed. In this part, solutions with different molar ratios of known analyte and internal standard will be prepared, analyzed by GC-MS, and their relative ratios can be used to construct a standard curve (technically, it should be a best fit line with high R^2 value).

After standard curves are constructed for each analyte to be assayed, the FAMEs in egg yolk can be quantified.

Procedure

Lab Period 1, Part 1: Construction of Standard Curves

1. Depending on the number of FAMEs that the class wants to analyze, students may be split into groups where each group will construct a standard curve for one of the analytes and then pool their data.
2. Prepare four solutions of 9-fluorenone and one known fatty acid methyl ester in hexanes. Weigh 18 mg of 9-fluorenone into a small vial or flask and 9 mg of desired methyl ester; add 5 mL of hexanes and swirl until

all the material is dissolved. Prepare a second solution with 18 mg of 9-fluorenone, 18 mg of methyl ester, third with 18 mg/27 mg, and 18 mg/36 mg, respectively. Record the masses to the maximum number of significant figures that is realistic for the balance used.

3. Calculate the exact number of moles of each compound, their molar ratios, and record them in the table provided (Table 8.1).

4. Inject a 2 μL sample into the GC-MS and record the spectrum. FAMEs will have a 4–8 min retention time, while 9-fluorenone has a retention time of about 9.5 min. These times may vary depending on the column and method used.

5. After all four samples have been run, record the areas under the peaks for the FAME and 9-fluorenone, calculate their ratio and plot them: molar ratio on one axis and area ratio on the other. Add a trend line and obtain the appropriate equation. This equation will be used to determine the concentration of FAMEs in the experimental sample.

Lab Period 2, Part 2: Transesterification of Phosphatidylcholine

1. Prepare a boiling water bath.

2. Carefully crack an egg and separate the yolk from the white. Place the yolk in a small beaker.

3. Equip a small vial with a stirrer bar and tare.

4. Carefully transfer approx. 100 mg of the yolk into the vial via a Pasteur pipette, and record the mass to the maximum number of significant figures realistic for the balance in use.

5. Add 3 mL of 2% methanolic H_2SO_4, tightly close the cap, and place in the boiling water bath. Incubate for 45 min with stirring.

6. Carefully take the vial out and place in the hood to slowly cool. *Do not* open the vial until it has cooled to room temperature as the contents are under pressure.

Table 8.1   Standard Curve for FAME v 9-Fluorenone

Vial #, 9-Fluo/FAME	FAME Used, MW							
	Exact Mass (mg)		# Moles		Molar Ratio: 9-Fluo/FAME	Peak Area		Area Ratio: 9-Fluo/FAME
	9-Fluo	FAME	9-Fluo	FAME		9-Fluo	FAME	
#1, 18 mg/9 mg								
#2, 18 mg/18 mg								
#3, 18 mg/27 mg								
#4, 18 mg/36 mg								

7. When cool, carefully open the vial and evaporate the liquid under a gentle stream of air. It will be faster if the vial is placed back into the water bath. Clamp securely to avoid tipping it over.
8. When dry, add 5 mL of saturated $NaHCO_3$ solution and set aside until all bubbling subsides.
9. Weigh 18 mg of 9-fluorenone into a clean small beaker or flask and dissolve in 3 mL of hexanes. Record the exact mass.
10. Transfer the hexane solution to the reaction mixture. Rinse the flask with two 1 mL portions of hexanes and add to the mixture. Close cap and shake for a minute or so.
11. Carefully open the cap (some gas may still be evolved) and transfer the contents (most likely an emulsion at this point) into a centrifuge vial via Pasteur pipette.
12. Centrifuge the sample for 10 min at 3000 rpm.
13. Carefully draw a few milliliters of the hexane layer (top) and place in a clean vial. Analyze by GC-MS.
14. Calculate the concentration of FAMEs by using the standard curves constructed in Part 1 and the known amount of 9-fluorenone placed in the reaction mixture. Report data in concentration of mg of FAME/mg of egg yolk.
15. Compare your results to the approximate reported composition of these acids: palmitic—26%, stearic—9%, oleic—44%, linoleic—11%.[2] If you are using a major brand of eggs, sometimes the manufacturer will provide this data.

References

1. Schulz, E.; Pugh, M. E. *J. Chem. Educ.* **2001**, *78*, 944–946.
2. Kaplan, M. E.; Simmons, E. R.; Hawkins, J. C.; Ruane, L. G.; Carney, J. M. *J. Sci. Food Agric.* **2015**, *95*(12), 2528–2532.

Chapter 9

Green Chemistry: Solventless Sequential Aldol and Michael Reactions in the Synthesis of Kröhnke Pyridine

Keywords: Macroscale, green chemistry, solventless reaction, tandem one pot synthesis, FTIR, NMR spectroscopy

Safety: Sodium hydroxide and acetic acid are corrosive and will cause severe burns on contact with skin. Acetophenone, benzaldehyde, and 2-acetylpyridine may cause irritation of eyes and airways. They are also flammable. For full MSDS, please visit www.sigmaaldrich.com.

Equipment Required

- Balance
- Mortar and pestle
- Büchner funnel and filter flask
- 250 mL round-bottomed flask
- Oil bath
- Thermometer
- Stirrer hotplate
- 2 mL syringe and needle
- FTIR spectrometer with dry KBr
- NMR spectrometer and sample tubes

Materials Required

- Sodium hydroxide (solid)
- Acetophenone
- Benzaldehyde (freshly distilled)
- 2-acetylpyridine
- Ammonium acetate
- Acetic acid
- Deuterochloroform

Purpose: To employ a solventless procedure to synthesize Kröhnke pyridine via one-pot tandem aldol and Michael reactions (Scheme 9.1).

Overall Reaction

This is a straightforward reaction employing classic carbonyl chemistry—aldol and Michael reactions. However, there are several important lessons to extract from the experiment: this procedure is solventless, tandem, one-pot, and high yielding when compared with other methods.

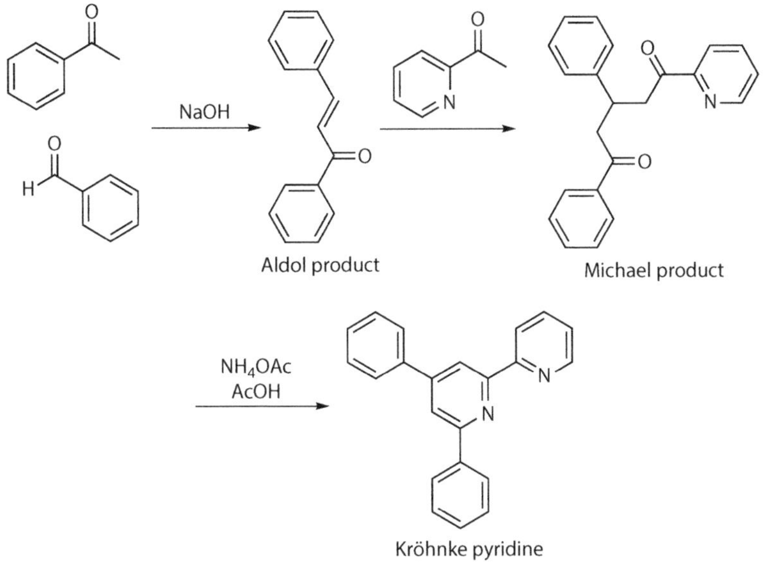

Scheme 9.1   **Synthetic scheme to prepare Kröhnke pyridine.**

One of the goals of Green Chemistry is to minimize toxic waste as much as possible. Developing solventless (or highly concentrated reactions) methods definitely fulfills this aim as most organic solvents are volatile and harmful, as well as flammable. Lack of solvent also simplifies isolation of the desired products and increases yield.

The fact that the reaction can proceed in the same reaction vessel without the need to isolate or purify the intermediate products makes this method more efficient. Also, purification often requires the use of solvents and transfers to other vessels, which inevitably results in a decrease in overall yield due to losses in transfer. Also, time is saved by the experimenter and can be spent on running more reactions, increasing output.

Finally, this more efficient and environmentally friendly method boasts higher yields than the original one reported in 1976.[1]

Needless to say, Green Chemistry faces a lot of challenges, but more and more recent articles feature words like "green," "tandem," "one-pot," "room temperature," and "facile."

When scaled to industrial levels, all these conditions make chemistry safer, more environmentally friendly, and economical.

Hence, this simple experiment is designed to introduce the student to highly efficient and "green" methods, and convince them of their superiority.

Procedure

1. Mix 1.1 mL (9.42 mmol) of acetophenone and 0.45 g (9.42 mmol) of finely ground NaOH in a dry mortar. Gently mix with a pestle for a couple of minutes. The color should change.
2. Add 0.96 mL (9.42 mmol) of benzaldehyde to the reaction mixture. Grind the reaction mixture for 10–15 min until the powder appears yellow (this is aldol product). Save a small sample for IR and NMR spectroscopy. Prepare a KBR disc for IR spectroscopy and record the NMR spectrum in $CDCl_3$ solution.
3. To the reaction mixture, add 1.05 mL (9.42 mmol) of 2-acetylpyridine and grind for an additional 10 min. Formation of the Michael product is indicated by the reaction mixture turning pink. Save a small sample for IR and NMR spectroscopy. Prepare a KBr disc for IR spectroscopy and record the NMR spectrum as a solution in $CDCl_3$.
4. Add a stirrer bar, 1.5 g (19.5 mmol) of ammonium acetate and 100 mL of acetic acid to a 250 mL round-bottomed flask, then transfer the Michael product from the mortar. Attach a reflux condenser (no water required) and heat under reflux for 2 h with stirring. Exercise caution not to overheat the reaction mixture (adjust heat so the "reflux ring" stays in the lower 1/3 segment of the reflux condenser).

5. Remove the mixture from heat and allow to cool to room temperature. Add 20 mL of distilled water to precipitate the product. Collect the product by filtration of the reaction mixture and dry under reduced pressure. Prepare a KBr disc for IR spectroscopy and record the NMR spectrum as a solution in $CDCl_3$.

Characterization[1]:

Aldol product: FTIR (KBr): ν 1600 (C=C), 1745 (C=O) cm^{-1}. ^{1}H NMR (300 MHz, $CDCl_3$): δ 8.02 (d, J = 7.0 Hz, 2H), 7.81 (d, J = 15.7 Hz, 1H), 7.63 (m, 2H), 7.52 (m, 4H), 7.40 (m, 3H).

Michael product: FTIR (KBr): ν 1695 (C=O) cm^{-1}. ^{1}H NMR (300 MHz, $CDCl_3$): δ 8.55 (m, 1H), 7.88 (m, 1H), 7.83 (m, 2H), 7.69 (dt, J = 7.65, J = 1.73 Hz, 1H), 7.44 (m, 1H), 7.32 (m, 3H), 7.25 (m, 2H), 7.20 (t, J = 7.48 Hz, 2H), 7.06, (m, 1H), 4.05 (m, 1H), 3.63 (m, 2H), 3.33 (m, 2H).

Kröhnke Pyridine: ν 705 (pyridine), 745 (phenyl) cm^{-1}. ^{1}H NMR (300 MHz, $CDCl_3$): δ 8.63 (m, 2H), 8.57 (d, J = 1.54 Hz, 1H), 8.13 (d, J = 6.95 Hz, 2H), 7.91 (d, J = 1.54 Hz, 1H), 7.75 (m, 3H), 7.42 (m, 6H), 7.26 (m, 1H).

References

1. Cave, G. W.; Raston, C. L. *J. Chem. Ed.* **2005**, *82*, 468–469.

Chapter 10

Carbonyl Chemistry in a Multistep Synthesis

Keywords: Carbonyl chemistry, multistep, tandem, one-pot synthesis, column chromatography, NMR spectroscopy

Safety: Sodium and potassium hydroxide and sulfuric and hydrochloric acids are corrosive and will cause severe burns on contact with skin. Diethyl ether, acetone, hexanes, ethyl acetate, and benzene are flammable solvents and irritants; they should be used in a well-ventilated area. Also, ensure that ether is freshly opened as it tends to form explosive peroxides upon prolonged contact with oxygen. Diethyl malonate is an irritant and is flammable. *p*-Toluenesulfonic acid monohydrate is an inhalation irritant. All reagents must be used in hoods and gloves should be worn. It should *never* be used outside a fume hood. Many labs prohibit it completely because of its long-term toxicity. For full MSDS, please visit www.sigmaaldrich.com.

Equipment Required

- Balance
- Stirrer hotplate
- Heating mantle with Variac or equivalent

- 25, 50, and 100 mL round-bottomed flasks with stirrer bars
- Rubber septa
- Reflux condenser (to fit 100 mL flask)
- Dish or large beaker for ice bath
- pH paper (pH 3–8 range or universal)
- 125 mL separatory funnel
- 125 mL Erlenmeyer flask
- Filter flask with Büchner funnel
- Pasteur pipettes
- 5 mL Plastic syringes with needles
- Small glass column or a 50 mL burette
- UV lamp
- Thirty 10 mL test tubes and rack
- TLC plates
- Desiccator
- Rotatory evaporator
- Supply of dry inert gas
- NMR spectrometer and sample tubes

Materials Required

- Ethanol (absolute)
- Sodium ethoxide (21% wt solution in ethanol)
- Acetone
- Isobutyraldehyde (2-methylpropanal)
- Diethyl ether
- $NaSO_4$, anhydrous
- NaOH aqueous solution, 2.5 M
- KOH aqueous solution, 3.8 M
- Diethyl malonate
- *p*-toluenesulfonic acid monohydrate
- Benzene, anhydrous
- 10% HCl, aqueous
- 20% H_2SO_4, aqueous
- $NaHCO_3$, saturated aqueous solution

- Brine
- Dichloromethane
- Ethyl acetate
- Hexanes
- Silica gel
- Deuterochloroform

Purpose: To practice multistep synthesis using classic carbonyl chemistry.

Overall Reaction

This lab employs classic carbonyl chemistry from reactions of enolates to the decarboxylation of β-ketoesters (Scheme 10.1). You should be able to draw arrow-pushing mechanisms for each of these steps. However, a deeper purpose of this chapter on carbonyl chemistry is to teach you an appreciation for multistep synthesis. One of the main challenges is overall yield. As we know, practically no reaction has quantitative yield: products are lost during each transfer, extraction, purification, or simply due to incomplete reaction, or the formation of by-products, in a particular process. That said, chemists constantly improve the methodology by optimizing reaction

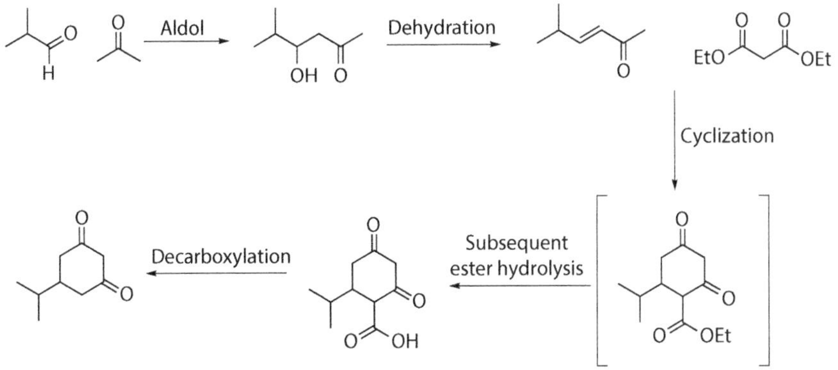

Scheme 10.1   Overall synthesis scheme.

conditions to minimize by-products and by improving isolation and purification techniques. We must make sure that an intermediate product has indeed been formed, before proceeding to the next step. If reaction intermediates are carried over without purification, it may be challenging to identify the desired product among many impurities.

Let's look at each step separately.

1. Aldol condensation

   Aldol condensation under basic conditions relies on the formation of an enolate anion by deprotonation of positions α to the carbonyl group (Scheme 10.2). Acetone will be deprotonated by NaOH and will attack the carbonyl carbon of isobutyraldehyde. Notice that isobutyraldehyde also has an α-hydrogen that could be removed by deprotonation. This would result in the formation of a mixture of products. In order to avoid the formation of unwanted aldol products, the order of addition is important: acetone is mixed with NaOH for some time to ensure complete formation of the enolate. Only then is isobutyraldehyde slowly added to that mixture. The desired cross-reaction is facilitated because the deprotonation site on the aldehyde is rather hindered, and the aldehyde is inherently more electrophilic than the ketone.

   Dehydration of aldol products (β-hydroxyketones) is very common and proceeds under acidic or basic conditions at slightly elevated temperature. As a result, an α,β-unsaturated ketone (conjugated 4π-system) is formed.

2. Cyclization and ester hydrolysis

   According to the mechanism proposed in Scheme 10.3, cyclization occurs by a Michael addition followed by an

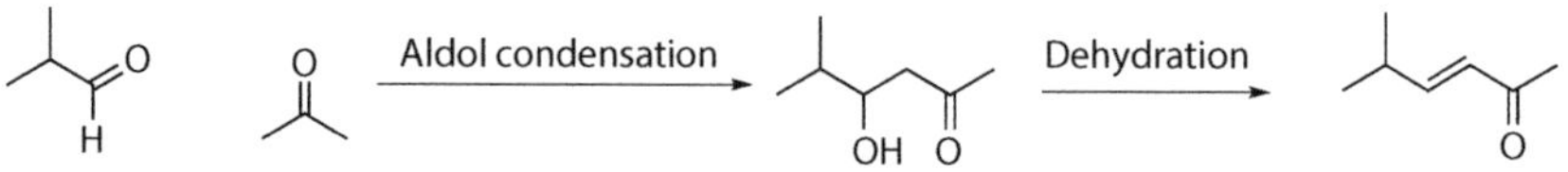

Scheme 10.2 Aldol condensation and dehydration.

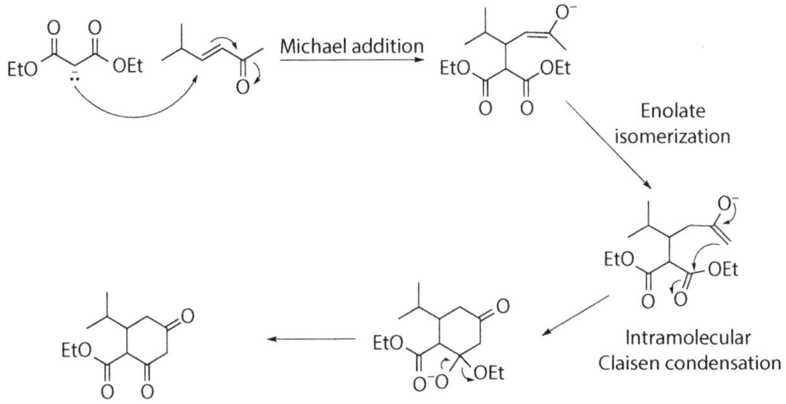

Scheme 10.3   Proposed mechanism of cyclization.

intramolecular Claisen condensation. Once again, notice the order of addition of the reagents: diethyl malonate is deprotonated first to form a stabilized enolate that will then attack the α,β-unsaturated ketone at the β-position, resulting in a 1,4-conjugate addition.

Suppose the order of addition were reversed: Would the reaction work in the same way simply starting with Claisen and continuing with Michael? In order to answer that question, compare the pK$_a$ of the α-hydrogen of diethyl malonate (9) with that of a typical ketone (19–20).

Subsequent ester hydrolysis (Scheme 10.4) is a common occurrence under both basic (saponification) and acidic conditions (reverse Fischer esterification).

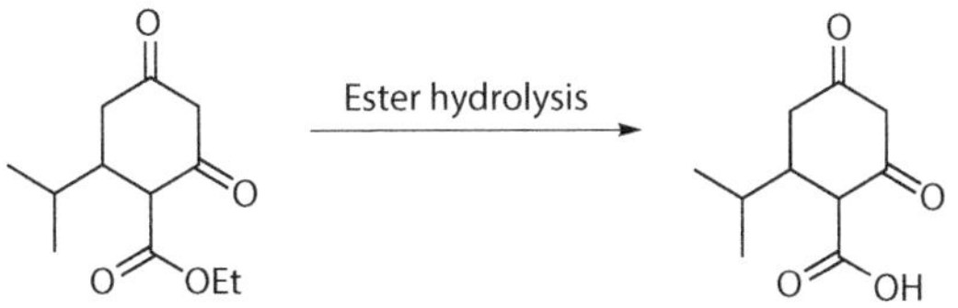

Scheme 10.4   Ester hydrolysis.

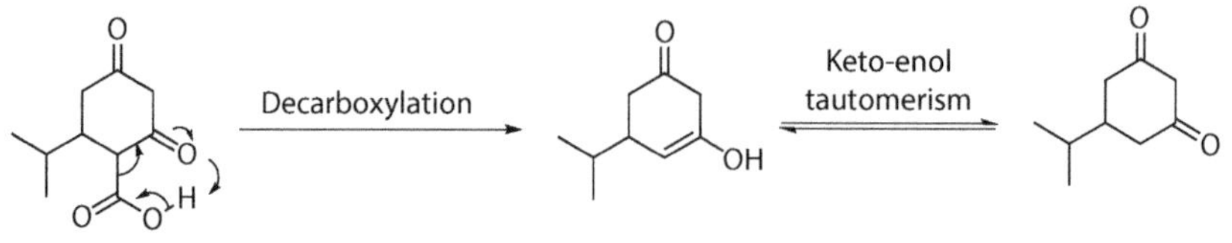

Scheme 10.5   Proposed decarboxylation mechanism.

3. Decarboxylation of β-ketoacids
   Decarboxylation of β-ketoacids occurs at slightly
   elevated temperatures (Scheme 10.5). In this case, a
   β-oxocarboxylic acid is undergoing decarboxylation. The
   mechanism is concerted and intramolecular, with a six-
   membered ring transition state.

Procedure

Lab Period 1, Part 1: Aldol Addition

1. Add 2.5 mL of acetone (34.0 mmol) and 0.93 mL of 2.5M
   NaOH (2.33 mmol) to a 50 mL round-bottomed flask
   with stirrer bar. Stir for 15 min at room temperature
   (Scheme 10.6).
2. Cool the reaction mixture to 0°C in an ice bath.
3. Dissolve 1.24 mL of isobutyraldehyde (13.6 mmol) in 2.5 mL
   of acetone (34.0 mmol) and add dropwise to the flask.
4. Cap the flask with a septum and allow the reaction
   to warm to room temperature and stir for 1 h. Place a
   syringe needle through the septum to release any pressure that may build in the flask.

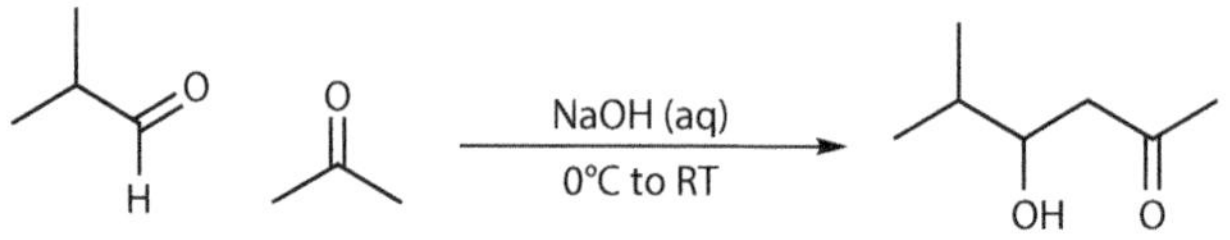

Scheme 10.6   Aldol addition.

5. After 1 h, quench the solution with 10% HCl. Add HCl dropwise until the pH reaches pH 7 (use litmus or universal indicator paper).
6. Transfer the reaction mixture to a separatory funnel. Rinse the flask with 10 mL of diethyl ether and add to the funnel.
7. Add 10 mL of water to the funnel and gently swirl.
8. Allow the layers to separate and drain the lower aqueous layer. Transfer the organic layer into a dry 125 mL Erlenmeyer flask.
9. Wash the aqueous layer twice more with 10 mL portions of diethyl ether and combine the organic layers.
10. Transfer the organic layer back into the separatory funnel and wash with 10 mL of brine.
11. Separate the layers and dry the organic layer over anhydrous $NaSO_4$.
12. Tare a 100 mL round-bottomed flask and carefully decant the ether layer into it. Wash the remaining $NaSO_4$ with 10 mL diethyl ether and combine the organic layers.
13. Concentrate the mixture under reduced pressure and record the mass of the product. The product should be a clear oil. Do not leave the product under reduced pressure too long as it is quite volatile.
14. Record the NMR spectrum. Some solvent peaks may still be present in the spectrum.

Lab Period 2, Part 2: Dehydration

1. Equip a 25 mL round-bottomed flask with a stirrer bar and add 0.25 g of *p*-toluenesulfonic acid monohydrate (1.31 mmol) and 4.0 g of anhydrous $NaSO_4$ (Scheme 10.7).

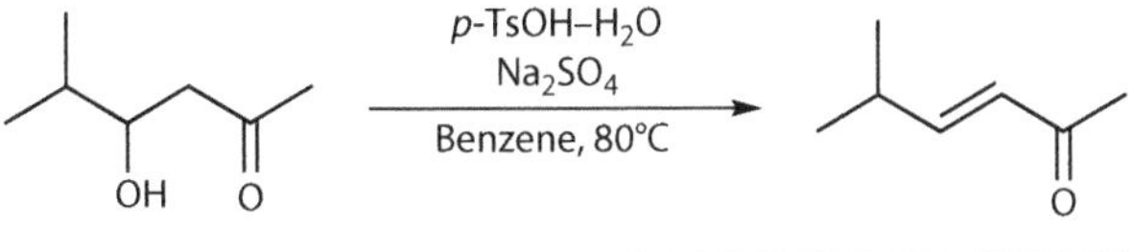

Scheme 10.7   Dehydration of the β-hydroxyketone.

2. Dissolve the aldol product from the previous lab period in 3 mL of dry benzene and add dropwise into reaction vessel. Wash the flask containing the aldol product with 3 more mL of benzene and add to the reaction flask.

3. Attach a reflux condenser and heat the mixture under reflux, with stirring, for 2 h.

4. Monitor the progress of the reaction by TLC starting at 1.5 h mark. Stain with anisaldehyde. When developed with 25% ethyl acetate in hexanes, the reported R_f values for the aldol starting material and the product are, respectively, $R_f = 0.3$ and $R_f = 0.7$.

5. Once all the starting material is gone or after 2.5 h of heating under reflux, allow the flask to cool to room temperature.

6. Quench the reaction mixture by dropwise addition of aqueous $NaHCO_3$ until the pH is 7.

7. Transfer the reaction mixture to a separatory funnel. Rinse the flask with 10 mL of diethyl ether and add to the funnel.

8. Add 10 mL of water to the funnel and gently swirl.

9. Allow the layers to separate and then drain the lower aqueous layer. Transfer the organic layer into a dry 125 mL Erlenmeyer flask.

10. Wash the aqueous layer twice more with 10 mL portions of diethyl ether and combine the organic layers.

11. Transfer the organic layer back into the separatory funnel and wash with 10 mL of brine.

12. Separate the layers and dry the organic layer over anhydrous Na_2SO_4.

13. Tare a 100 mL round-bottomed flask and carefully decant the ether layer, then wash the remaining Na_2SO_4 with 10 mL diethyl ether and combine the organic layers.

14. Concentrate the mixture under reduced pressure. The product should be a clear oil. Do not leave the product under reduced pressure too long as it is rather volatile.

15. Record the NMR spectrum. Solvent peaks may still be present in the spectrum.

Lab Period 3, Part 3: Cyclization and Ester Hydrolysis

1. Flame-dry (or bake in an oven for a few hours) a 100 mL round-bottomed flask, quickly transfer it to desiccator, and allow to cool to room temperature (Scheme 10.8).
2. Equip the flask with a stirrer bar and place a septum over it. Supply a positive pressure of inert gas (nitrogen or argon) through a needle. Place another needle to serve as an outlet valve.
3. Add 15 mL of absolute ethanol and 3.5 mL of the 21% NaOEt solution (9.4 mmol).
4. Turn on the stirring and add 1.4 mL diethyl malonate dropwise (9.3 mmol).
5. Turn on the heat and replace the septum with a reflux condenser when the flask is warm (just a minute or so; do not let it get too hot).
6. Heat the reaction mixture under reflux, with stirring, for 15 min.
7. Dissolve the dehydration product in 3 mL of absolute ethanol and add dropwise through the reflux condenser. Wash the flask with an additional 2 mL of absolute ethanol and add to the reaction vessel. *If you prepare this solution in advance, cap the flask with a septum to keep atmospheric moisture out.*
8. Heat under reflux for an additional 30 min.

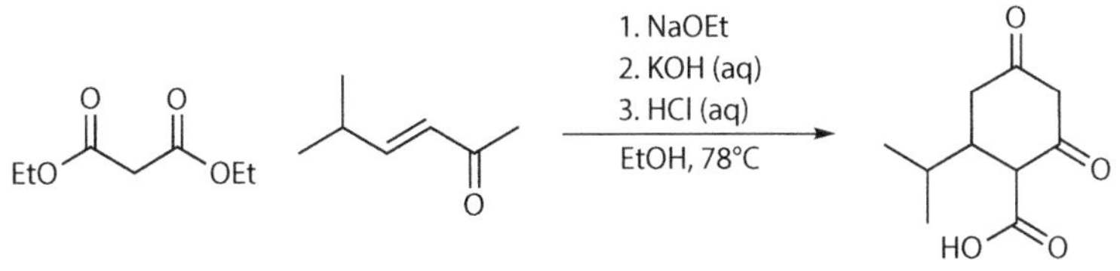

Scheme 10.8 Cyclization and ester eydrolysis.

9. Add 5.1 mL of 3.8 M KOH (19.4 mmol) dropwise through the condenser and heat under reflux for an additional 2 h.
10. Remove the reaction from heat, allow it to cool, and add 10% HCl dropwise until the pH is 7.
11. Concentrate the reaction mixture under reduced pressure and securely store the crude mixture until next lab period.

Lab Period 4, Part 4: Decarboxylation and Purification of the Final Product

1. Equip the flask containing the product from the previous section with a reflux condenser, turn on the stirring, and heat the reaction mixture to a gentle reflux (Scheme 10.9).
2. Add 6 mL of 20% H_2SO_4 through the condenser and continue heating under reflux for 1 h or until gas evolution has ceased.
3. Allow the reaction mixture to cool to room temperature and bring the pH to 4 by dropwise addition of $NaHCO_3$ solution.
4. Transfer the reaction mixture to a separatory funnel. Rinse the flask with 10 mL of dichloromethane and add to the funnel.
5. Allow the layers to separate. Drain the lower organic layer (this time it is on the bottom as DCM is denser than water) into a dry 125 mL Erlenmeyer flask.

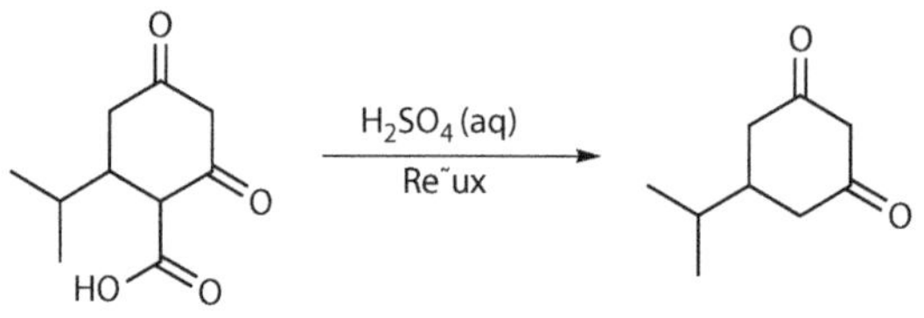

Scheme 10.9 Decarboxylation of β-oxocarboxylic acid.

6. Wash the aqueous layer twice more with 10 mL portions of dichloromethane and combine the organic layers.

7. Transfer the organic layer back into the separatory funnel and wash with 10 mL of brine.

8. Separate the layers and dry the organic layer over anhydrous Na_2SO_4.

9. Tare a 100 mL round-bottomed flask and carefully decant the ether layer, then wash the remaining Na_2SO_4 with 10 mL dichloromethane and combine the organic layers.

10. Concentrate the mixture under reduced pressure. The product should be a brown oil.

11. Dissolve a small amount of the crude oil in 1 mL of dichloromethane and spot a TLC plate. Develop in 75% ethyl acetate in hexanes and visualize under UV light. The product R_f is reported as 0.3 under these conditions.

12. Prepare a slurry of 50 g of silica in 75:25 ethyl acetate:hexanes and pack a small column (~2.5 cm in diameter).

13. Dissolve the crude product mixture in a minimum amount of dichloromethane and load the column.

14. Elute with 75:25 ethyl acetate:hexanes and collect 30 fractions (7–10 mL each). Ramp solvent polarity by switching to pure ethyl acetate and allow last five fractions be eluted with ethyl acetate. Depending on the column parameters, the product should elute between fractions 14 and 24. Confirm this by TLC.

15. Combine the fractions containing the product and concentrate under reduced pressure.

16. Dissolve the product in a small amount of dichloromethane, transfer into a smaller flask or vial, and concentrate under reduced pressure. You may also want to place the product under high vacuum to remove the final traces of the solvent.

17. Record the NMR spectrum.

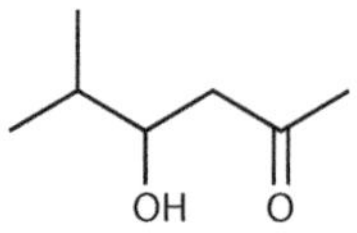

Figure 10.1   Aldol product.

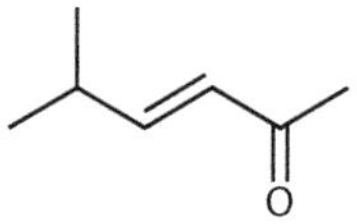

Figure 10.2   Dehydration product.

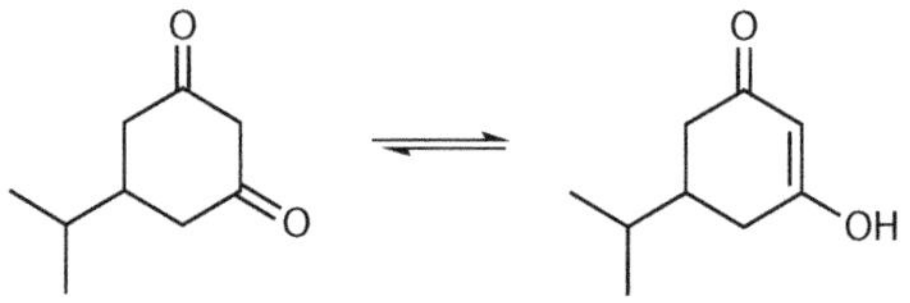

Figure 10.3   Cyclization/decarboxylation (final) product.

Characterization[1]:

^{1}H NMR (CDCl$_3$) δ: 3.76–3.82 (m, 1H), 2.96 (bs, 1H), 2.44–2.60 (m, 2H), 2.14 (s, 3H), 1.58–1.66 (m, 1H), 0.84–0.92 (m, 6H) (Figure 10.1).*

^{1}H NMR (CDCl$_3$) δ: 6.72 (dd, 1H), 5.99 (d, 1H), 2.40–2.46 (m, 1H), 2.12 (s, 3H), 1.03 (d, 6H) (Figure 10.2).

^{1}H NMR (CDCl$_3$) δ: 5.42 (s, 1H), 3.38 (d, 2H), 2.68 (dd, 2H), 2.40 (dd, 2H), 1.82–1.90 (m, 1H), 1.58–1.68 (m, 1H), 0.96 (d, 6H) (Figure 10.3).

References

1. Duff, D. B.; Abbe, T. G.; Goess, B. C. *J. Chem. Educ.* **2012**, *89*, 406–408.

---

* The crude product may contain other peaks, such as impurities and solvent.

Chapter 11

Multistep Synthesis of a Bioactive Peptidomimetic

Keywords: Bioactive compound, LC–MS, TLC, one-pot multi-step synthesis, solid-phase support

Safety: Trifluoroacetic acid is extremely corrosive and will cause severe burns. It is also an inhalation hazard. In addition to goggles, wear gloves. Bromoacetic acid is also corrosive and should be handled with care. Amines will cause irritation and burns upon direct contact. For full MSDS, please visit www.sigmaaldrich.com.

Equipment Required

- Analytical balance
- 5 mL polypropylene syringe with a porous polypropylene frit
- Needles
- 50–100 μL syringe

- Caps for syringes
- Capillary tubes
- 10 microcentrifuge tubes or small test tubes and rack
- TLC plate and developing chamber
- Pasteur pipettes
- Mass spectrometer coupled with LC
- Vials for LC–MS
- UV lamp

Materials Required

- *N,N*-dimethylformamide (DMF)
- Dichloromethane (DCM)
- Bromoacetic acid
- Piperidine
- *N,N'*-diisopropylcarbodiimide (DIC)
- Acetaldehyde
- *p*-Chloranil (2,3,5,6-tetrachloro1,4-benzoquinone)
- Trifluoroacetic acid (TFA)
- Formic acid
- Rink amide resin
- Acetonitrile
- Tetrahydrofurfurylamine
- 3,3-diphenylpropylamine

Stock Solutions

- 20% (v/v) piperidine in DMF
- 1.0M DIC in DMF
- 1.2M 2-bromoacetic in DMF
- 1.0M diphenylpropylamine in DMF
- 1.0M tetrahydrofurfurylamine in DMF
- 2% (v/v) acetaldehyde in DMF. *Must be prepared fresh.*
- 2% (w/v) chloranil in DMF. *Must be prepared fresh.*

- 10% (v/v) TFA in water
- 10% methanol in DCM for TLC

Purpose: To synthesize a bioactive peptidomimetic.

Overall Reaction

Organic synthesis is widely used in the biomedical field: this lab features the synthesis of a bioactive peptidomimetic (Scheme 11.1). This tripeptide was found to have potential anti-cancer properties.[1] The reason it is called a peptidomimetic is because it resembles peptides found naturally (in this case it is repeating *N*-substituted glycine units). However, substitution at the amide nitrogen makes this molecule unnatural.

This multistep synthesis takes place in one reaction vessel by reiterating cycles of acylation and displacement. Three fragments will be joined to form a tripeptide during this lab, but the reaction sequence could continue and longer peptides could be obtained.

The process can be broken down into three major stages: deprotection of a solid support resin, synthesis of the peptide, and cleavage of the peptide from solid support.

Scheme 11.2 depicts the removal of an F-moc protecting group to expose a free amine group enabling the reaction to start. Piperidine is used as an organic base to perform the deprotection. This is one of the reasons that F-moc is so popular as a protecting group as it can be removed under mild conditions.

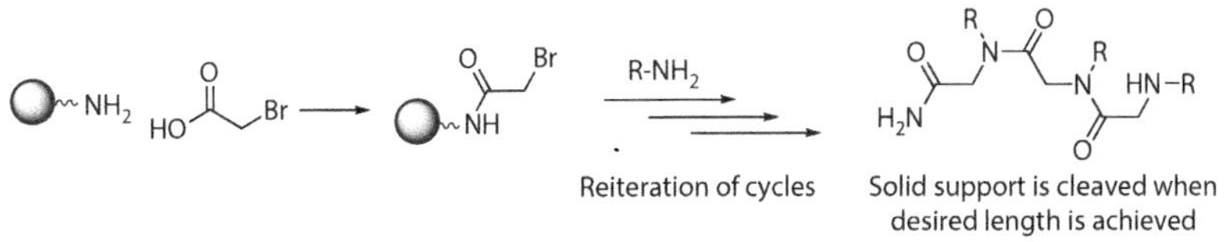

Reiteration of cycles   Solid support is cleaved when desired length is achieved

Scheme 11.1 Synthetic path to a peptidomimetic.

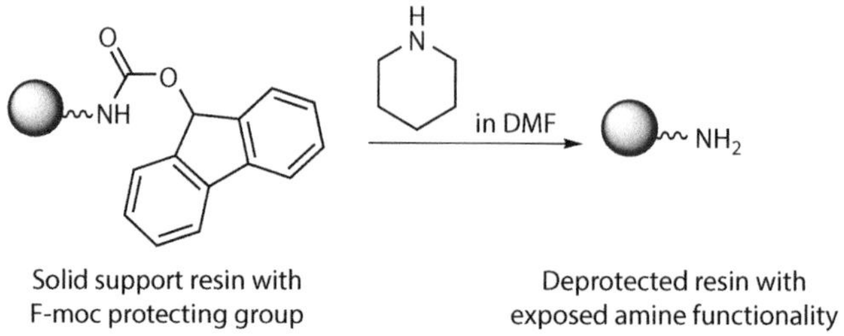

Solid support resin with
F-moc protecting group

Deprotected resin with
exposed amine functionality

Scheme 11.2   Cleavage of an F-moc protecting group.

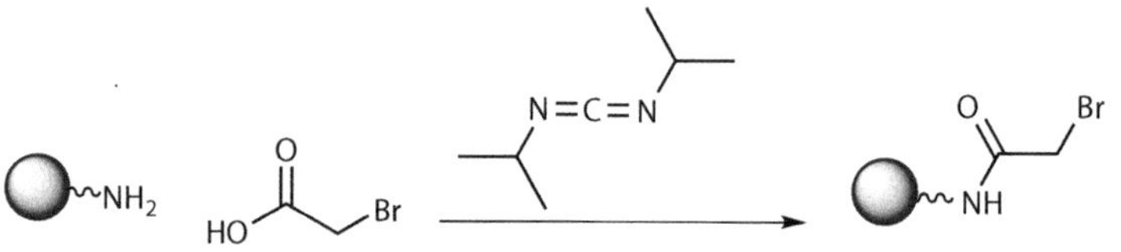

Scheme 11.3   Formation of a peptide bond with DIC.

After deprotection is complete, the free amine group can be reacted with 2-bromoacetic acid to form a peptide bond (Scheme 11.3). A coupling reagent (*N*,*N'*-diisopropylcarbodiimide, DIC) is necessary to complete this transformation (review formation of peptide bonds). Its mechanism is identical to that using *N*,*N'*-dicyclohexylcarbodiimide (DCC), a coupling reagent often featured in undergraduate texts.

The next step involves an S_N2 attack on the carbon-atom-bearing bromine. This displacement results in the formation of a carbon–nitrogen bond (Scheme 11.4). At this point, the

Scheme 11.4   Nucleophilic displacement of bromide by a nitrogen nucleophile.

cycle is complete and a new amide bond can be made with 2-bromoacetic acid, followed by a new displacement, until the desired length of the peptide is achieved. Amine nucleophiles can have different substituents. In this lab, diphenylpropylamine will be used for the first two cycles followed by tetrahydrofurfurylamine.

Chloranil tests will be performed through this reaction to ensure the completeness of each step. Chloranil turns blue in the presence of secondary amines. You will be watching for either the absence or presence of blue color depending on the step.

The final step is to release the newly synthesized peptide from its solid support (Scheme 11.5). This reaction is achieved using trifluoroacetic acid (TFA). At this point, the product can be collected, dried, and analyzed.

Figure 11.1 shows a tripeptide that can be synthesized by sequential addition of 2-bromoacetic acid, diphenylpropylamine (×2) and tetrahydrofurfurylamine, followed by cleavage from the solid support.

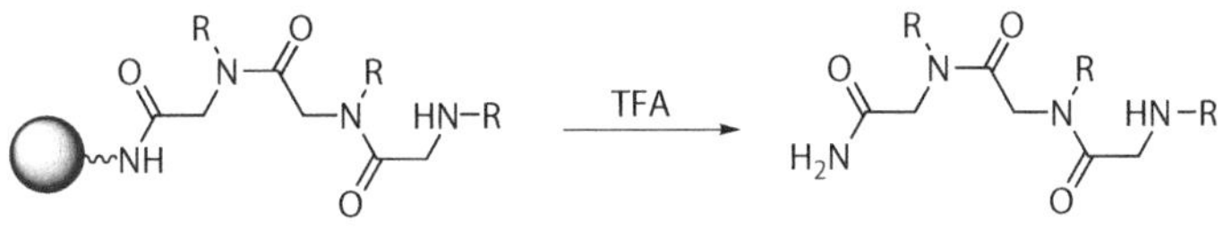

Scheme 11.5 **Cleavage of the tripeptide from the solid support.**

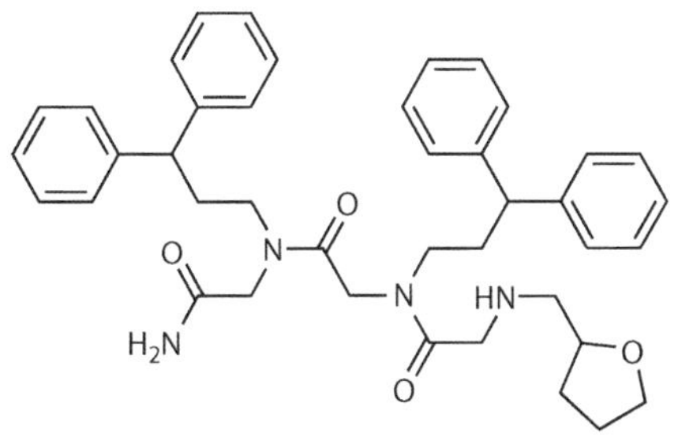

Figure 11.1 **Peptidomimetic synthesized by the procedure described.**

Procedure

1. Weigh 50 mg of F-moc-protected rink amide resin and place it into a polypropylene syringe.
2. Measure out 1 mL of DMF and draw it into a syringe.
3. Cap the syringe and periodically agitate for 10 min until resin swells.
4. Meanwhile, perform control chloranil tests. Prepare three small tubes and combine 20 µL of 2% acetaldehyde solution with 20 µL of chloranil solution. Add a drop of piperidine solution to the first one, diphenylpropylamine to the second one, and tetrahydrofurfurylamine solutions to the third. Observe the colors of the mixtures. The piperidine solution should turn blue (secondary amine), while the other three should remain clear (primary amines). If you observe different results, consult with your instructor.
5. Wash the resin three times with 3 mL portions of 20% piperidine solution. This step will cleave F-moc group and expose the free amine.
6. Wash the resin three times with 1 mL portions of DMF, allowing 1 min for each wash. Discard the wash.
7. Draw 0.5 mL of 1.0M DIC solution and 0.5 mL of 1.2M 2-bromoacetic acid solution and agitate for 2 min. This step will couple the free amine end of the resin with a carboxylic acid forming a peptide bond.
8. Wash the resin three times with 1 mL portions of DMF, allowing 1 min for each wash. Discard the wash.
9. The chloranil test: combine 20 µL of 2% acetaldehyde solution with 20 µL of chloranil solution in a small test tube, vial or microcentrifuge tube. Add a few resin beads from the reactor. Be careful not to spill or add too much as it will result in product loss. At this point, the solution should be a clear yellow as there are no secondary amines present.
10. Draw 0.5 mL of 1.0M diphenylpropylamine solution and agitate occasionally for 5 min.

11. Wash the resin three times with 1 mL portions of DMF, allowing 1 min for each wash.
12. Perform the chloranil test. The solution should now turn blue due to the presence of a secondary amine.
13. Repeat steps 7–12. At this point, you should have synthesized a dimer with two diphenylpropylamine units.
14. Repeat steps 7–12 *except step 10*. Draw 0.5 mL of 1.0M tetrahydrofurfurylamine solution instead.
15. Wash the resin three times with 1 mL portions of dichloromethane. DCM is very volatile and may squirt out of syringe due to pressure buildup. Also, do not leave DCM in the syringe for a long period of time as the plastic will dissolve.
16. Draw 1 mL of 10% TFA solution into syringe, cap it, and gently agitate for 10 min. *Exercise extreme caution: TFA is very corrosive!* This step deprotects the resin and frees the peptoid. The product is now in solution.
17. Depress the plunger gently and collect solution in a small test tube.
18. Pull a TLC spotter from a pipette and spot a TLC plate. Run the product sample in 10:90 MeOH:DCM and visualize under a UV lamp. Compare the R_f value of the product to the reported value.
19. Prepare a sample for LC–MS analysis.

Characterization[2]:

Method for ESI LC–MS product analysis: dilute the final product in 50% acetonitrile-H_2O. Inject 0.5 µL at 400 µL/min flow with 85% acetonitrile-H_2O with 0.1% formic acid. Run in positive mode for 3 min with 2 min of equilibration time.

TLC analysis: eluent used 10:90 methanol:DCM. Visualize under UV lamp.

Monomer (diphenylpropylamine) $R_f = 0.41$

Dipeptide (two diphenylpropylamine groups) $R_f = 0.69$

Tripeptide $R_f = 0.65$

References

1. Mas-Moruno, C.; Cruz, L. J.; Mora, P.; Francesch, A.; Messeguer, A.; Perez-Paya, E.; Albericio, F. *J. Med. Chem.* **2007**, *50*, 2443–2449.
2. Utku, Y.; Rohatgi, A.; Yoo, B.; Kirshenbaum, K.; Zuckermann, R. N.; Pohl, N. L. *J. Chem. Ed.* **2010**, *87*, 637–639.

Chapter 12

Total Synthesis of a Natural Product: Caffeic Acid Phenethyl Ester (CAPE)

Keywords: Total synthesis, multistep, natural product, TLC, NMR

Safety: Acetate anhydride and thionyl chloride are very reactive with water and will cause severe burns on contact. Thionyl chloride will produce toxic HCl gas on contact with water. Handle those reagents in the hood and wear gloves in addition to goggles. Pyridine is toxic and will cause irritation. Ethyl acetate, hexanes, toluene, methanol, and ethanol are toxic, flammable solvents. Dichloromethane is volatile, irritant, and carcinogen. For full MSDS, please visit www.sigmaaldrich.com.

Equipment Required

- Analytical balance
- 25, 50, 100, and 250 mL round-bottomed flasks with stirrer bars

- Condenser to fit 25 mL round-bottomed flask
- Septa
- Small syringe needle
- 250 µL syringe
- 50, 125, and 250 mL Erlenmeyer flasks
- Stirrer hotplate
- Ice bath
- Büchner funnel
- Watch glass
- Filter flask
- 125 mL separatory funnel
- Oven
- Melting point apparatus with capillaries
- 10–20 mL glass syringe
- UV lamp and/or $KMnO_4$ stain/heat gun
- TLC plates and developing chamber
- Silica gel
- Column
- 20 × 10 mL test tubes and rack
- Pasteur pipettes
- Rotatory evaporator
- NMR spectrometer

Materials Required

- Caffeic acid
- NaOH, 1M, aqueous
- Acetate anhydride
- Ethanol, 95%
- Dichloromethane (DCM)
- *N,N*-dimethylformamide (DMF)
- Thionyl chloride
- Dry toluene
- Dry pyridine

- Ethyl acetate
- Brine
- Sodium bicarbonate, aqueous solution
- Hexanes
- $MgSO_4$, anhydrous
- Methanol
- K_2CO_3

Purpose: To synthesize a natural product from readily available materials.

Overall Reaction

Many natural products possess medicinal properties and many synthetic chemists are devising efficient methods for the synthesis of such products in their labs. The importance of doing so is based on the fact that frequently only very small quantities of natural product can be isolated. Furthermore, natural sources could be limited and overextraction could lead to environmental damage. The art of synthesizing a natural product from simple and widely available precursors is called *total synthesis.* It is indeed an art, as oftentimes, natural products contain complicated scaffolds with specific stereocenters. It is not uncommon for a few years to pass between discovery and structure elucidation of a natural product and its first successful total synthesis.

The time scale of this lab does not allow for total synthesis of complex molecules but it does not mean it cannot be done. Caffeic acid phenethyl ester is a simple natural product found in propolis from honeybees[1] and is said to have inhibitory effects on lipooxygenase-5 activity,[2] an enzyme responsible for metabolizing arachidonic acid.[3] Polyphenols, such as CAPE, have been found to have valuable medicinal properties such as vasoprotection, anticancer, anti-inflammatory, and anti-obesity effects.[4] Touaibia and coworkers have devised a facile synthesis of CAPE and adapted it to an undergraduate laboratory setting.[5]

This experiment features total synthesis of a biologically active material and employs all the important techniques that you have learned, such as TLC monitoring, intermediate isolation and purification, and product analysis.

The overall synthesis (Scheme 12.1) can be broken down into three major parts: acetylation of the two phenol groups of caffeic acid, esterification of the carboxylic acid group of caffeic acid, and deprotection of the acetates to regenerate free phenols.

Part 1 is easily achieved by reacting the phenol groups with acetate anhydride (Scheme 12.2). The reaction is very similar to the synthesis of aspirin from acetylsalicylic acid. This step

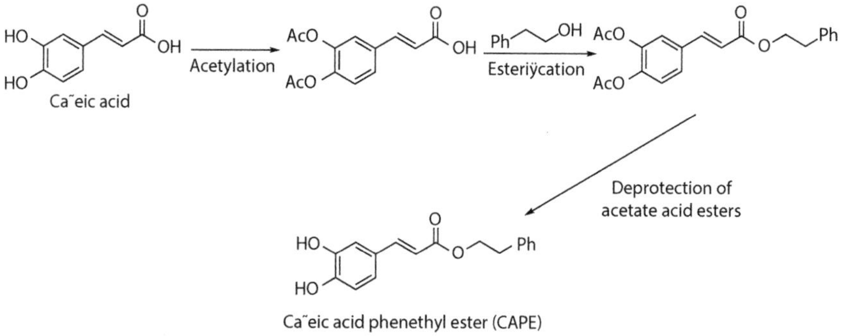

Scheme 12.1   Synthesis of CAPE.

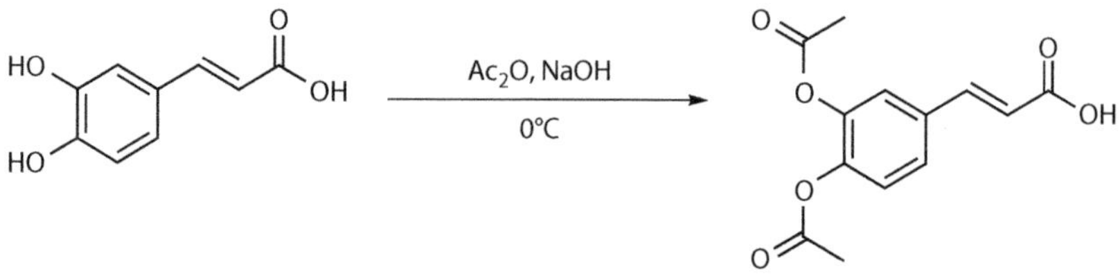

Scheme 12.2   Protection of the phenol groups as acetate esters.

is necessary as acetate esters will mask phenols and prevent them from reacting with the later acyl chloride intermediate, hence avoiding by-products.

Part 2 involves converting the carboxylic acid functional group into the acyl chloride; acyl chloride is an excellent electrophile and chloride is a good leaving group. This will allow for efficient esterification with phenethyl alcohol (Scheme 12.3).

Finally, the acetate esters will be deprotected with a base (saponified) and CAPE will be synthesized. Interestingly, such conditions selectively deprotect phenyl acetates and not the phenethyl ester (Scheme 12.4). Propose a rationale for difference in reactivity.

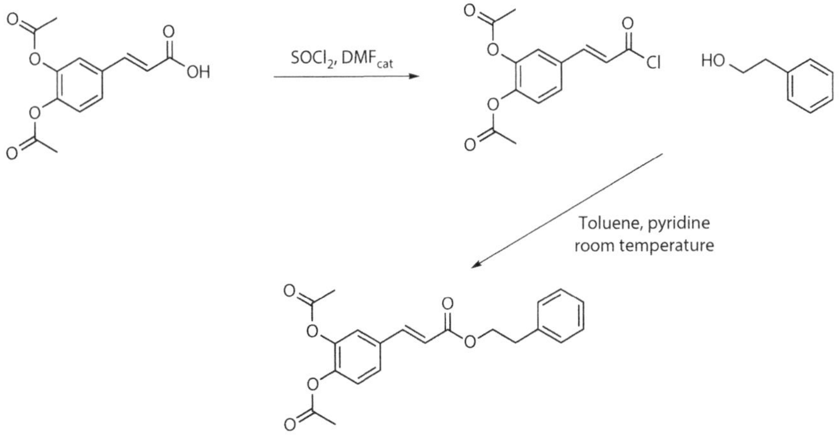

Scheme 12.3 Formation of acyl halide and subsequent esterification.

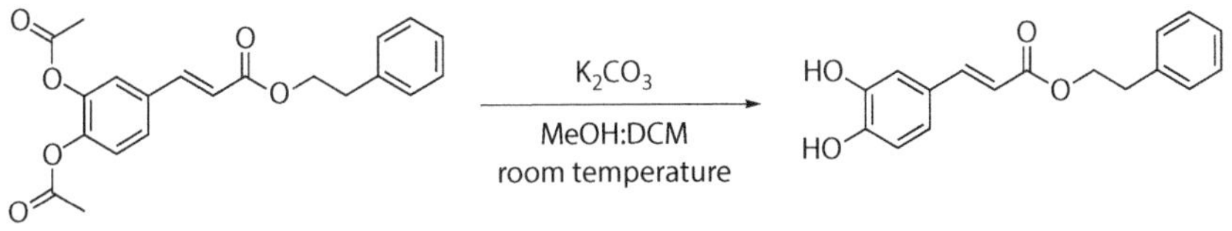

Scheme 12.4 Deprotection of acetate esters.

Procedure

Lab Period 1: Protection of the Phenol Groups with Acetate Anhydride

1. Equip a 50 mL round-bottomed flask with a stirrer bar.
2. Weigh 1 g of caffeic acid (5.5 mmol) and combine it with 15 mL of 1M NaOH in the flask. Stir until all the solid dissolves.
3. Place the reaction mixture into an ice bath and slowly add 2 mL of acetate anhydride with stirring. Allow the reaction to proceed for 30 min in an ice bath.
4. Collect the crystals formed using a Büchner funnel, and wash the product with a few milliliters of chilled water.
5. Place the crystals in a 50 mL Erlenmeyer flask and add 15 mL of ethanol. Gently heat to boiling with occasional stirring until all product dissolves. Addition of a little more ethanol may be necessary if the solid does not go into solution after starting to boil.
6. Remove the flask from heat and allow to cool to room temperature. Then chill in an ice bath for 10 min.
7. Collect the crystals by suction filtration, place them on a watch glass, and dry in an oven for 15 min at 100°C.
8. Record the melting point and record an NMR spectrum. Compare with reference values.
9. Transfer the recrystallized product into a beaker, cover loosely with foil or parafilm, and store until the next lab period.

Lab Period 2: Formation of Acyl Chloride and Esterification with Phenethyl Alcohol

1. Equip a flame-dried 25 mL round-bottomed flask with a stirrer bar and condenser.
2. Weigh 250 mg of diacetyl caffeic acid (0.94 mmol) and transfer into the flask.

3. With a glass syringe, slowly add 8 mL of thionyl chloride and two drops of DMF.

4. Heat the reaction mixture under reflux, with stirring, for 1 h.

5. Allow the reaction to cool to room temperature and concentrate under reduced pressure. *Thionyl chloride will damage the rotovap pump. It is strongly recommended that you set up a cold trap and fill it with aqueous sodium bicarbonate solution to neutralize any acidic vapors. Make sure that pump exhaust is in the hood.*

6. When evaporated to dryness, dissolve the residue in 10 mL of dry toluene.

7. Dropwise, add 1 mL of dry pyridine.

8. Dropwise, add 125 µL of 2-phenylethanol (1.04 mmol).

9. Cap the flask with a septum, insert a syringe needle through the septum, and stir until next lab period at room temperature. At least 4 h of reaction time is required for reasonable yields.

Lab Period 3: Isolation and Purification of Acetylated CAPE

1. Concentrate the reaction mixture under reduced pressure, dissolve the residue in 10 mL of ethyl acetate, and transfer it into a 125 mL separatory funnel.

2. Rinse the flask with two 10 mL portions of ethyl acetate and combine the washings in the separatory funnel.

3. Wash the ethyl acetate layer twice with 20 mL of water and once with 20 mL of brine.

4. Transfer the organic layer into a 125 mL Erlenmeyer flask and dry over $MgSO_4$ for 5 min.

5. Decant the organic layer into a 100 mL round-bottomed flask, wash the $MgSO_4$ residue with 10 mL of ethyl acetate, and decant the washing into the flask.

6. Spot a TLC plate with this solution and run with 20:80 ethyl acetate:hexanes. Develop with a $KMnO_4$ stain or visualize under UV light.

7. Concentrate the reaction solution under reduced pressure.

8. Prepare a slurry of 30 g of silica in 10:90 mixture of ethyl acetate:hexanes and pack the column. Do not forget to add a layer of sand on top.

9. Dissolve the crude sample in a small amount of DCM and load the column.

10. Elute the column with 50 mL 10:90 mixture followed by 50 mL of 20:80 mixture of ethyl acetate:hexanes.

11. Determine which fractions contain the product (see Appendix), combine those fractions in a 250 mL round-bottomed flask and concentrate under reduced pressure.

12. Record the NMR spectrum and the melting point of the product.

Lab Period 4: Deprotection of Phenols and Purification of CAPE

1. Place 200 mg of acetylated CAPE (0.54 mmol) into a 10 mL flask with a stirrer bar.

2. Add 5 mL of a 1:1 MeOH:DCM mixture and 225 mg of K_2CO_3 and stir at room temperature for 15 min.

3. After 15 min, start monitoring the reaction by TLC. Run the plate in a 50:50 mixture of ethyl acetate:hexanes.

4. When all starting material has disappeared by TLC, concentrate the mixture under reduced pressure.

5. Dissolve the residue in 10 mL of ethyl acetate and transfer into a 125 mL separatory funnel.

6. Rinse the flask with two 10 mL portions of ethyl acetate and combine the washings in the separatory funnel.

7. Wash the ethyl acetate layer twice with 20 mL of water and once with 20 mL of brine.

8. Transfer the organic layer into a 125 mL Erlenmeyer flask and dry over $MgSO_4$ for 5 min.

9. Decant the organic layer into a 100 mL round-bottomed flask, wash the $MgSO_4$ residue with 10 mL of ethyl acetate, and decant the liquid into the flask.

10. Concentrate the solution under reduced pressure.
11. Dissolve the product in a few milliliters of DCM and transfer into a tared small flask or vial. Rinse the 100 mL round-bottomed flask twice with 1–2 mL portions of DCM and add to the tared flask.
12. Evaporate the DCM to dryness and record the mass. It is best to allow the product to stay under vacuum for a few minutes.
13. Record the NMR spectrum and the melting point of the sample.

Characterization:

Diacetylcaffeic acid[5]: white solid, mp 190°C–192°C. ^{1}H NMR (200 MHz, DMSO-d$_6$): δ 7.7 (br s, 2H), 7.55 (d, 1H, J = 18.6 Hz), 7.2 (d, 1H, J = 8.6 Hz), 6.5 (d, 1H, J = 18.6 Hz), 2.3 (s, 6H) (Figure 12.1).

Acetylated CAPE[5]: white solid, mp 80°C–82°C. ^{1}H NMR (200 MHz, CDCl$_3$): δ 7.6 (d, 1H, J = 18.6 Hz), 7.5–7.3 (m, 8H), 6.35 (d, 1H, J = 18.6 Hz), 4.3 (t, 2H, J = 4.7 Hz), 3.1 (t, 2H, J = 4.7 Hz), 2.3 (s, 6H). R_f = 0.26 in 20:80 ethyl acetate:hexanes (Figure 12.2).

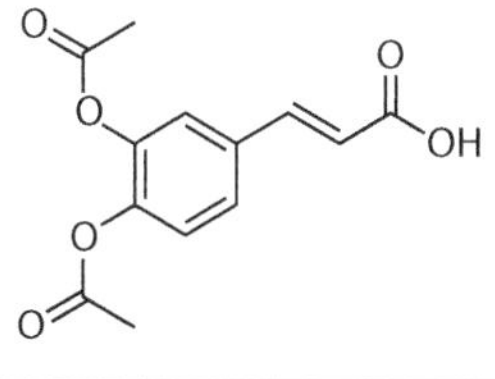

Figure 12.1   Diacetylcaffeic acid.

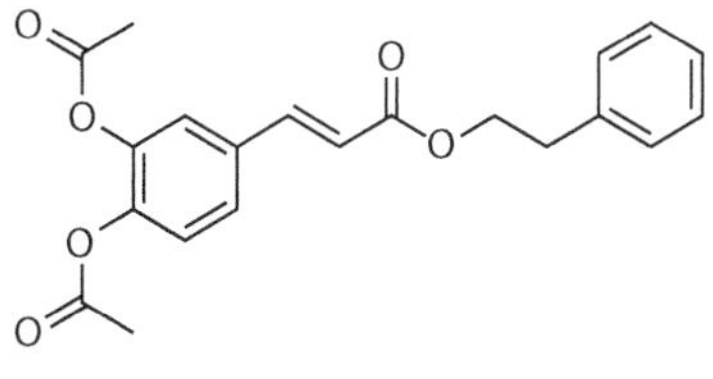

Figure 12.2   Acetylated CAPE.

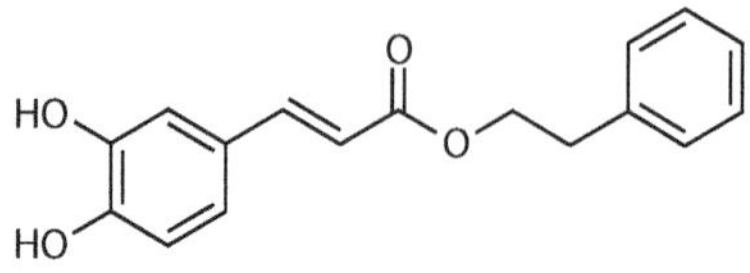

Figure 12.3 CAPE.

***CAPE*[5]:** white solid, mp[1,5] 126°C–128°C. ^{1}H NMR (200 MHz, C_2D_6O): δ 8.4 (br s, 2H), 7.5 (d, 1H, J = 18.6 Hz), 7.4–7.2 (m, 6H), 7.1 (d, 1H, J = 8.6 Hz), 6.9 (d, 1H, J = 8.6 Hz), 6.3 (d, 1H, J = 8.6 Hz), 4.4 (t, 2H, J = 4.7 Hz), 3.0 (t, 2H, J = 4.7 Hz) (Figure 12.3).

References

1. Grunberger, D.; Banerjee, R.; Eisinger, K.; Oltz, E. M.; Efros, L.; Caldwell, M.; Estevez, V.; Nakanishi, K. *Experientia* **1988**, *44*, 230–232.
2. Mirzoeva, O. K.; Calder, P. C. *Prostaglandins Leukot. Essent. Fatty Acids* **1996**, *55*, 441–449.
3. Peters-Golden, M.; Henderson, W. R. *N. Engl. J. Med.* **2007**, *357*, 1841–1854.
4. Williamson, G.; Manach, C. *Am. J. Clin. Nutr.* **2005**, *81*, 243S–255S.
5. Touaibia, M.; Guay, M. *J. Chem. Educ.* **2011**, *88*, 473–475.

Appendix: Essential Laboratory Techniques

Choosing Glassware

Every piece of glassware has its purpose. Flasks are usually used to carry out reactions, test tubes to collect column fractions or set up a series of test reactions, etc. However, most importantly, when choosing glassware, size is an issue. The volume of a reaction flask should be no less than twice the volume of the reaction (and quenching solutions). In other words, try to keep a reaction vessel 1/4 full, 1/3—at max. The headspace will allow for quenching solutions to be added and will accommodate any bubbling, if the reaction is heated or evolves gas.

Drying Glassware

In many organic experiments, water is an unwelcome molecule. Even miniscule amounts of it can produce sizable amounts of by-products or prevent a reaction from happening all together. Flame-drying glassware is one technique that keeps atmospheric water away. In order to do that, one needs to either flame-dry a piece of glassware until all signs of condensation are gone (a Bunsen burner will suffice) or bake it overnight in an oven (at least 150°C). Most importantly, to prevent water from recondensing, this piece of glassware needs to be cooled in a desiccator and, once room temperature is reached, put under a blanket of inert, dry gas.

Purging with Inert Gas

Often, reactions are sensitive to moisture, or air, or both. In that case, it is necessary to exclude these elements by purging the reaction vessel with a dry, inert gas, such as nitrogen or argon. A flame-dried and cooled flask is quickly capped with a septum, and a needle with a positive pressure of inert gas is inserted. Then another needle is inserted through the septum. The second needle serves as an outlet, and as nitrogen or argon fills the flask, the air is replaced. Continue purging for 2–3 min (Figure A.1). After that, leave the needle with the gas in as it will provide a positive pressure of inert gas.

Figure A.1   Supply of dry nitrogen gas and a pressure release needle.

Setting a Pressure Release Valve

Even when reactions are not sensitive to atmosphere, it is still a good idea to keep the reaction vessel capped. Reasons include keeping debris away and acting as a safeguard against spills. Most importantly, many organic solvents are volatile and will evaporate after a few hours of reaction time. The best way to keep the reaction flask closed is with a septum. However, if the reaction is known to evolve gas or is heated, pressure will build inside. To circumvent this problem, insert a syringe needle through the septum. That will be enough to equilibrate pressures and prevent evaporation.

Handling of Pyrophoric and Water-Sensitive Reagents

Make sure that all surfaces coming in contact with these reagents are dry and there are no open sources of water or flammable reagents. Weigh out in small quantities, usually no more than a gram at a time.

Setting Up a Chromatography Column

Column chromatography is a widely used method for compound separation. In order to properly pack the column, the thing that is most important to keep in mind is that silica must be uniformly packed with no air bubbles. In order to achieve that, measure out the silica into a dry Erlenmeyer flask and add enough eluting mixture to make a slurry (Figure A.2).

Keep agitating it until the suspension is uniform (Figure A.3).

Then, quickly pour the slurry into a column (extra rinses may be needed). At this point, the column valve must be open at all times, so place a large flask under the column to collect runoff solvent (Figure A.4).

Figure A.2   Dry silica in a flask.

Figure A.3   Making a silica slurry.

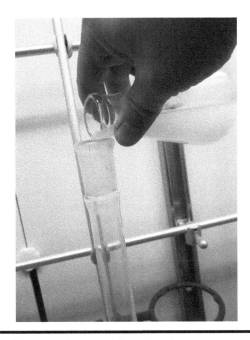

Figure A.4 Packing a column with silica slurry.

Also, never allow the column to go dry—always keep enough solvent above the silica layer. Once the silica settles, take a cork ring and tap the column gently, but firmly (Figure A.5). This will further settle the silica and expel any trapped air bubbles.

With a few centimeters of solvent covering the silica, add a pinch of sand (Figure A.6). Sand does not play any role in the separation but will protect the top silica layer from being disturbed in subsequent solvent additions.

Additional solvent can be supplied by fitting a solvent bulb or rigging a separatory funnel on top of the column (Figure A.7).

Again, remember to always keep the column running and do not let it go dry. Now it is ready to be loaded with the mixture to be separated. Dissolve the crude in the smallest amount of eluting solution and gently transfer to the column when the solvent layer is about to touch the top of the silica. Allow for the solution to settle, then very gently add a few

Figure A.5   Aiding silica slurry to settle uniformly.

Figure A.6   Protective layer of sand.

Figure A.7   Adding solvent through separatory funnel.

milliliters of eluting solution and let it settle. Gently top off the column and start collecting fractions in test tubes. A thin layer chromatography (TLC) of the crude mixture will give an idea of when the product will elute.

Most compounds are colorless and cannot be seen when eluting from a column. In order to find them, collect a few fractions and spot a TLC plate sequentially with a drop of each fraction. Then develop without running it. This will give you an idea of which fractions contain the product. To identify the compound, spot a TLC plate with a spotter, run, develop, and compare against a TLC of the crude material.

Pulling Spotters

This technique involves heating a Pasteur pipette or a capillary tube in a flame until soft, and then pulling it. By doing so, not only does the overall length increase, but the diameter

Figure A.8   Heating a Pasteur pipette.

decreases, allowing for very fine tubes. Such small diameter tubes are useful for spotting TLC plates without overloading them and keeping the spot small.

Begin by heating the Pasteur pipette at the widening on Bunsen burner (Figure A.8).

Rotate it in the flame until the glass becomes soft and pliable (Figure A.9).

Just a fraction of a second before you are ready to pull, remove it from the flame and then pull in a smooth but quick motion. One pipette can be pulled to several feet of spotters (Figure A.10).

This technique will take some practice, so be patient. And remember, hot glass looks exactly the same as cold (Figure A.11).

Spotting and Running a TLC Plate

When monitoring reactions by TLC, small, accurate spots provide best results. Pull some spotters from a pipette, obtain a TLC plate, and mark a line about 1 cm off the bottom.

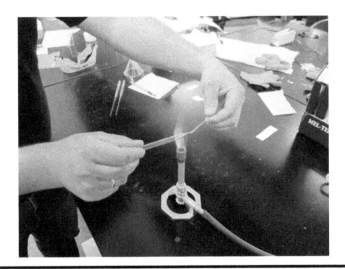

Figure A.9   Heat pipette until soft and pliable.

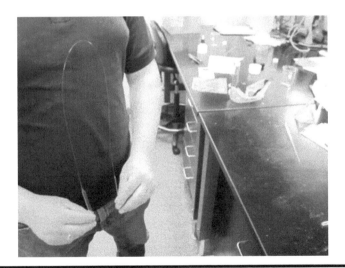

Figure A.10   Pulled pipette.

Always use a pencil and mark very lightly so as not to damage the silica layer. Now mark light dots or crosses on that line not closer than 0.5 cm from each other (as you gain experience, you will be able to fit more)—this is where the solution in question will be loaded. Take a spotter and dip in a solution to be tested, then gently touch the TLC plate. Let the spot

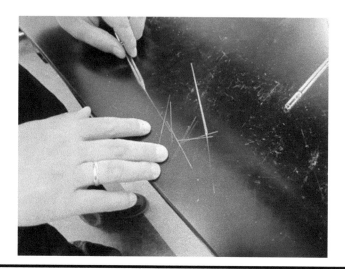

Figure A.11   Spotters broken up and ready to use.

dry and reapply 2–3 times. Try to keep the spot as small as possible.

Then, place the TLC plate in a developing chamber or a beaker with eluting solution and cover with a watch glass or tin foil. Ensure that the solvent layer is below the marked line. Allow the solvent to rise (via capillary action) and almost reach the top. Take the plate out and quickly mark the solvent line with a pencil. Visualize spots either under UV or develop in a special solution. The method will depend on the functional groups in the molecules being analyzed.

A TLC plate will not only provide information about the identity of the compound, but also can be used to monitor the reaction. R_f values are also useful in predicting retention times/volumes when running a column.

Dry Solvents and Reagents

Many research labs are equipped with solvent stills, which supply dry solvents of choice. However, in an undergraduate lab, this may just mean opening a fresh bottle of high-purity solvent. Regardless, care should be exercised in keeping those solvents dry: Minimize their exposure to the atmosphere and always dispense in flame-dried glassware, cooled in a desiccator. When these solvents need to be stored for prolonged periods of time, keep them under a blanket of dry, inert gas and add some activated sieves. Cap the flask with a lightly greased glass stopper (septa are OK for short storage periods but can be compromised by many organics after prolonged exposure).

Setting Up an Apparatus

Some reactions involve several pieces of glassware connected to each other. A distillation apparatus is the most common but the same principles apply to similar setups. To begin building the apparatus, place a heating mantle or a stirrer hot plate on a ring about 10 cm off the surface. This will allow cutting off heat fast (by lowering it) in the case reaction boils over (Figure A.12).

Next, secure the reaction flask firmly by the neck. Allow enough space to accommodate a heating mantle (pictured) or a stirrer hot plate (Figure A.13).

These are all the clamps you *need*! Keck clamps are recommended to secure condensers or adapters but are not a must. When assembled properly, gravity will hold pieces in place as long as the foundation is straight and true.

Do not forget to grease lightly around glass joints and firmly attach water hoses. If heating to a substantially elevated temperature, the use of glass wool or metal foil for insulation will significantly speed the process (Figure A.14).

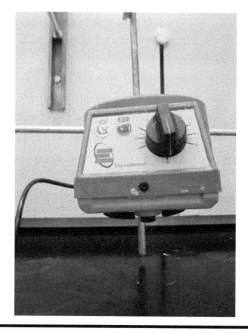

Figure A.12 Heating mantle on a ring.

Figure A.13 Reaction flask must be firmly clamped.

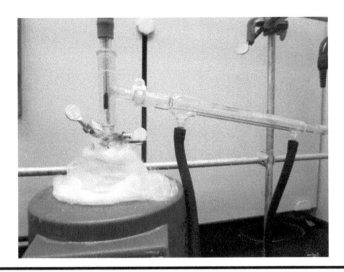

Figure A.14   Complete short-path distillation setup.

Index

T - #0950 - 101024 - C0 - 234/156/6 - PB - 9781482244960 - Gloss Lamination